T0332168

Integer Programming

Mathematics and Its Applications (*East European Series*)

Volume 46

Stanisław Walukiewicz

System Research Institute,
Polish Academy of Sciences, Warszawa

Integer Programming

KLUWER ACADEMIC PUBLISHERS
DORDRECHT / BOSTON / LONDON

PWN — POLISH SCIENTIFIC PUBLISHERS
WARSZAWA

Library of Congress Cataloging-in-Publication Data

Walukiewicz, Stanisław.
    Integer programming / Stanisław Walukiewicz.
        p. cm. — (Mathematics and its applications / East European series: v. 46)
        "Revised translation from the Polish original Programowanie dyskretne, published in 1986"
    T.p. verso.
        ISBN 0-7923-0726-7
        1. Integer programming.   I. Walukiewicz, Stanisław.
    Programowanie dyskretne.   II. Title.   III. Series: Mathematics and its applications
    (Kluwer Academic Publishers. East European series)
    T57.74.W37   1990
    519.7′7—dc20

ISBN 0-7923-0726-7

90—4332
CIP

Published by PWN—Polish Scientific Publishers,
Miodowa 10, 00-251 Warszawa, Poland
in co-edition with Kluwer Academic Publishers,
P.O. Box 17, 3300 AA Dordrecht, The Netherlands

Distributors for the U.S.A. and Canada:
Kluwer Academic Publishers,
101 Philip Drive, Norwell, MA 02061, U.S.A.

Distributors for Albania, Bulgaria, People's Republic of China, Cuba, Czechoslovakia, Hungary,
Korean People's Republic, Mongolia, Poland, Romania, the U.S.S.R., Vietnam and Yugoslavia:
ARS POLONA,
Krakowskie Przedmieście 7, 00-068, Warszawa, Poland

Distributors for all remaining countries:
Kluwer Academic Publishers Group,
P.O. Box 322, 3300 AH Dordrecht, The Netherlands

Revised translation from the Polish original *Programowanie dyskretne*, published in 1986 by
Państwowe Wydawnictwo Naukowe, Warszawa (translated by the Author)

To My Mother —

Bronisława Walukiewicz

SERIES EDITOR'S PREFACE

'Et moi, ..., so j'avait su comment en revenir, je n'y serais point alle.'

Jules Verne

The series is divergent; therefore we may be able to do something with it.

O. Heaviside

One service mathematics has rendered the human race. It has put common sense back where it belongs, on the topmost shelf next to the dusty canister labelled 'discarded non-sense'.

Eric T. Bell

Mathematics is a tool for thought. A highly necessary tool in a world where both feedback and nonlinearities abound. Similarly, all kinds of parts of mathematics serve as tools for other parts and for other sciences.

Applying a simple rewriting rule to the quote on the right above one finds such statements as: 'One service topology has rendered mathematical physics ...'; 'One service logic has rendered computer science ...'; 'One service category theory has rendered mathematics ...'. All arguably true. And all statements obtainable this way form part of the raison d'être of this series.

This series, *Mathematics and Its Applications*, started in 1977. Now that over one hundred volumes have appeared it seems opportune to reexamine its scope. At the time I wrote

"Growing specialization and diversification have brought a host of monographs and textbooks on increasingly specialized topics. However, the 'tree' of knowledge of mathematics and related fields does not grow only by putting forth new branches. It also happens, quite often in fact, that branches which were thought to be completely disparate are suddenly seen to be related. Further, the kind and level of sophistication of mathematics applied in various sciences has changed drastically in recent years: measure theory is used (non-trivially) in regional and theoretical economics; algebraic geometry interacts with physics; the Minkowski lemma, coding theory and the structure of water meet one another in packing and covering theory; quantum fields, crystal defects and mathematical programming profit from homotopy theory; Lie algebras are relevant to filtering; and prediction and electrical engineering can use Stein spaces. And in addition to this there are such new emerging subdisciplines as 'experimental mathematics', 'CFD', 'completely integrable systems', 'chaos, synergetics and large-scale order', which are almost impossible to fit into the existing classification schemes. They draw upon widely different sections of mathematics."

By and large, all this still applied today. It is still true that at first sight mathematics seems rather fragmented and that to find, see, and exploit the deeper underlying interrelations more effort is needed and so are books that can help mathematicians and scientists do so. Accordingly MIA will continue to try to make such books available.

If anything, the description I gave in 1977 is now an understatement. To the examples of interaction areas one should add string theory where Riemann surfaces, algebraic geometry, modular functions, knots, quantum field theory, Kac-Moody algebras, monstrous moonshine (and more) all come together. And to the examples of things which can be usefully applied let me add the topic 'finite geometry'; a combination of words which sounds like it might not even exist, let alone be applicable. And yet it is being applied: to statictics via designs, to radar/sonar detection arrays (via finite projective planes), and to bus connections of VLSI chips (via difference sets). There seems to be no part of (so-called pure) mathematics that is not in immediate danger of being applied. And, accordingly, the applied mathematician needs to be aware of much more. Besides analysis and numerics, the traditional workhorses, he may need all kinds of combinatorics, algebra, probability, and so on.

In addition, the applied scientist needs to cope increasingly with the nonlinear world and the extra mathematical sophistication that this requires. For that is where is where the rewards are. Linear models are honest and a bit sad and depressing: proportional efforts and results. It is in the nonlinear world that infinitesimal inputs may result in macroscopic outputs (or vice versa). To appreciate what I am hinting at: if electronics were linear, we would have no fun with transistors and computers; we would have no TV; in fact you would not be reading these lines.

There is also no safety in ignoring such outlandish things as nonstandard analysis, superspace and anticommuting integration, p-adic and ultrametric space. All three have applications in both electrical engineering and physics. Once, complex numbers were equally outlandish, but they frequently proved the shortest path between 'real' results. Similarly, the first two topics named have already provided a number of 'wormhole' paths. There is no telling where all this is leading-fortunately.

Thus the original scope of the series, which for various (sound) reasons now comprises five subseries: white (Japan), yellow (China), red (USSR), blue (Eastern Europe), and green (everything else), still applies. It has been enlarged a bit to include books treating of the tools from one subdiscipline which are used in others. Thus the series still aims at books dealing with:

— a central concept which plays an important role in several different mathematical and/or scientific specialization areas;

— new applications of the results and ideas from one area of scientific endeavour into another;

— influences which the results, problems and concepts of one field of enquiry have, and have had, on the development of another.

Euler wrote

> "For since the fabric of the universe is most perfect and the work of a most wise Creator, nothing at all takes place in the Universe in which some rule of maximum or minimum does not appear."

That sort of statement implies that there is nothing in mathematics, at least applied mathematics, that has not to do with optimization, i.e. programming. Add to that that much in this world (possibly all) is discrete and the place of (nonlinear and linear) integer programming seems assured in the general scheme of things. And so, of course, it is. Correspondingly, to being in principle applicable to everything, even linear integer programming, the subject of this book is very difficult ($\mathcal{NP}$-complete in the language of comutational complexity).

It is also a field in which a great deal is happening at the moment, and that seems likely to continue, and, finally, it seems to me be an area of research that is less well known than it should be to fellow mathematical specialists, and it is perhaps insufficiently appreciated in terms of the depth and beauty of its problems and results. (This may well hold for all of discrete mathematics, and, to a lesser extent, for algebra as well.)

Thus, I am happy to welcome this volume, suitable both for self-study and for courses, on the topic of linear integer programming.

The shortest path between two truths in the real domain passes through the complex domain.
J. Hadamard

Never lend books, for no one ever returns them; the only books I have in my library are books that other folk have lent me.
Anatole France

La physique ne nous donne pas seulement l'occasion de résoudre des problemes ... elle nous fait pressentir la solution.
H. Poincaré

The function of an expert is not to be more right then other people, but to be wrong for more sophisticated reasons.
David Butler

Bussum, September 1989   Michiel Hazewinkel

Preface

Integer programming originated in the mid-fifties as a branch of linear programming. Now it is a rapidly developing subject in itself, having many common points with operations research on one hand and with discrete mathematics on the other. A measurement of this development may be the increasing numbers of publications, reflected in the classified bibliographies edited by Kasting in 1976 (4704 items), Hausman in 1978 (3162 items) and by Randow in 1981 (3924 items). The number of applications of integer programming has also sharply increased in recent years.

The aim of this book is to present in a unified way the theory and methodology of integer programming including recent results. It contains examples how these results may be used in the construction of more efficient numerical methods. It is designed as a textbook for a one- or two-semester course at a department of operations research, management science or industrial administration. It may also be used as source of references.

In contrast to other books on integer programming, we consistently use here the notion of equivalence and relaxation: both are defined in Chapter 1. This makes all our considerations shorter and allows us to introduce simple notations.

The book has ten chapters. Each chapter is to a certain extent a closed unit, ended, with the exception of Chapter 10, with bibliographical notes and exercises. The chapters are divided into sections, some longer sections are divided into subsections. We use typical numbering of chapters, sections, figures and tables.

The main objective of integer programming is the construction of numerical methods for solving discrete optimization problems. First we describe the idea of a method, next discuss results on which it is based, and at the end we present a formal description of it in terms of a simple model language similar to ALGOL 60. After such a presentation it would not be difficult to write a computer program for the considered methods.

The book is organized in the following way.

In Chapter 1, we discuss the concept of and give formal definitions for an equivalence and relaxation. Next we present examples of applications of integer programming, describe ideas of different integer programming methods and of different ways of measuring their efficiency. In this chapter, we also introduce the notations which we will use further on.

For the sake of completeness, we review the main results of linear programming in Chapter 2. New here is a description of the ellipsoid method and discussion of subgradient methods used increasingly more often in integer programming.

In Chapter 3, we discuss some integer programming problems which may be solved as linear programming problems and give a description of cutting-plane methods.

The next four chapters are devoted to branch-and-bound methods as the most efficient approach to solving integer programming problems. In Chapter 4, we describe the idea of such an approach and the use of linear programming in it, while in Chapter 7, we consider other relaxations. The second part of Chapter 7 is devoted to duality in integer programming. Computational experiments show that it is worth reformulating a given integer programming problem before solving it. Different ways of reformulating a given problem are considered in Chapter 6. They are mainly based on the result of analysis of the knapsack problem done in Chapter 5. In Chapter 10, we give a general description of a system of computer codes in which the above-mentioned reformulation plays a very important role and in fact determines the efficiency of the system.

In Chapter 8, we consider problems with binary matrices. For such particular problems, similarly as for the knapsack problem, many results have been obtained and some of them are used in the construction of solution methods for general integer programming problems.

Chapter 9 is devoted to near-optimal methods and, in particular, to ones in which it is possible to estimate the error of such methods.

The book is an extension of my lectures given at the University of Copenhagen and at Warsaw University. Some parts of it were presented at seminars of the Mathematical Programming Department (MPD) of the System Research Institute. I wish to thank my friends from the MPD for their constructive criticism and in particular Dr. Ignacy Kaliszewski, who read the manuscript and proposed many improvements.

Stanisław Walukiewicz

Table of Contents

CHAPTER 1

Introduction

The aim of this chapter is to introduce the basic definitions and relations of integer programming and to give examples of its applications.

We present the notation used further on and discuss our understanding of such concepts as algorithms (methods), method efficiency, data structure, and so on.

This chapter gives also a more detailed description of the content of the book than found in the Preface.

1.1. FORMULATION OF THE INTEGER PROGRAMMING PROBLEM

Many questions in management and design may be formulated as optimization problems in which we are looking for such values of decision variables (decisions) for which the objective function takes an extreme value (maximal or minimal) subject to all of the constraints imposed on decision variables. More than 30 years of experience have proven usefulness of such an approach. In many practical problems, all or part of the decision variables take only integer values. We give examples of such problems in Section 1.5.

The main objective of *integer programming* is the construction of numerical methods for solving such problems, i.e., problems of integer programming.

In an *integer programming problem*, we are looking for a vector $x^* = (x_1^*, \ldots, x_n^*)$ $\in R^n$ such that an objective function

$$f(x) = \sum_{j=1}^{n} c_j x_j \tag{1.1}$$

takes the maximal value for $x = x^*$ subject to constraints

$$\sum_{j=1}^{n} a_{ij} x_j \leqslant b_i, \quad i = 1, \ldots, m, \tag{1.2}$$

$$x_j \geqslant 0, \quad j \in N = \{1, \ldots, n\}, \tag{1.3}$$

$$x_j \text{ integer}, \quad j \in N, \tag{1.4}$$

where R^n is the n-dimensional space of reals.

This is an example of a *linear integer programming* problem since all of its relations (1.1) and (1.2) are linear.

In the next section, we show that under weak assumptions, any nonlinear integer programming problem may be transformed into an equivalent linear problem.

In the above formulation, $f(x)$ is called an *objective function* while c_j is a *coefficient of an objective function*, $j = 1, ..., n$. The term *constraint* is traditionally reserved for relations (1.2), although (1.3) and (1.4) also restrict feasibility regions for (decision) variables x_j, $j = 1, ..., n$. Thus the considered problem has m constraints and n variables. Data a_{ij} is called a *coefficient of the ith constraint*, while b_i is its *right-hand side*.

The main assumption made in this book is based on the integrality of all data in problem (1.1)–(1.4), so $c_1, ..., c_n, a_{11}, ..., a_{mn}$ and $b_1, ..., b_m$ are integers. Such integrality of data is not really a restriction from a practical point of view since in any practical integer programming problem, all data are rational numbers and can be made integers after multiplication of (1.1) and (1.2) by a suitable integer. We will often use these assumptions which have a deep theoretical meaning and which will be discussed in Chapter 3.

Using the matrix notation, the problem (1.1)–(1.4), which we denote by P, may be written in a more compact form as

$$v(P) = \max \{cx | Ax \leqslant b, x \geqslant 0, x \in Z^n\}. \qquad (P)$$

Now c is an n-dimensional vector, i.e., $c = (c_1, ..., c_n)$, A is a matrix with m rows and n columns, i.e., $A = (a_{ij})$, $i = 1, ..., m$, $j = 1, ..., n$, while $b = (b_1, ..., b_m)$. We write $x \geqslant 0$ for $x_j \geqslant 0$, $j = 1, ..., n$. By Z^n we denote a set of all n-dimensional vectors with integral components. By $v(P)$ we denote the optimal (in our case, maximal) value of the objective function in P which we will call a *value of problem P*.

We will hereby denote by P a given optimization problem within a section, while in another (sub)section, P may denote another problem.

To simplify notation, we will not distinguish a vector from its component by typesetting, but in agreement with a mathematical programming tradition, we will denote a vector by x and by x_j its jth component. For the same reasons, everywhere where it does not lead to misunderstandings, we will not distinguish the transposition of vectors and matrices and will always assume that all scalar products are well-defined. For example, in the notation cx, c is an n-dimensional row vector and x is an n-dimensional column vector, similarly as in the notation Ax, while in a product xA, x is an m-dimensional row vector providing A has m rows and n columns.

A set of all $x \in R^n$ satisfying (1.2)–(1.4) is called a *feasible (solution) set* of P and will be denoted by $F(P)$. So

$$F(P) = \{x \in R^n | Ax \leqslant b, x \geqslant 0, x \in Z^n\}.$$

Each vector $x \in F(P)$ is called a *feasible solution* (*vector*, *element*) of problem P. If $F(P)$ is contained in some ball in R^n of a finite radius, then $F(P)$ is said to be *bounded*, otherwise we have an *unbounded feasible solution set*. If $F(P) = \emptyset$, i.e., the feasible solution set is empty, which means that constraints (1.2)–(1.4) are inconsistent, then P is said to be *inconsistent*. Then, obviously, an optimal solution x^* does not exist. If P models a practical problem, then it follows from $F(P) = \emptyset$ that a mistake must have been made in the modelling process since in practice some decisions that are taken may be nonoptimal ones, therefore $F(P) \neq \emptyset$.

Each vector $x^* \in F(P)$ such that

$$\infty > f(x^*) \geqslant f(x) \quad \text{for all } x \in F(P)$$

is called an *(optimal) solution* of P. So

$$v(P) = f(x^*) = \sum_{j=1}^{h} c_j x_j^*.$$

We write $v(P) < \infty$ to denote the fact that the objective function takes a finite value at any optimal point. If this is not the case, then P is called *unbounded* and we will write $v(P) = \infty$, in contrast to the *bounded problem*. If in a practical problem P, $v(P) = \infty$, then a mistake must have been made in the modelling process as in practical problems all decisions x_j are finite. Unboundness of $F(P)$ is only necessary, but not sufficient to have $v(P) = \infty$.

Let $F^*(P)$ be a set of all optimal solutions of P. Usually P involves finding one element of $F^*(P)$, but in many practical problems it is useful to know all or many optimal solutions. We will discuss this question in Chapter 10. For an inconsistent problem P, we obviously have $F^*(P) = \varnothing$.

An integer programming problem may be formulated in a more general way as

$$v(P) = \sup_{x \in F(P)} f(x), \tag{P}$$

with $F(P) \subseteq D$, where D is a given discrete set in R^n, i.e., such a set that for each $x \in D$ there exists an $\varepsilon > 0$ such that in the ε-neighbourhood there are no other points of D. (An ε-neighbourhood is a ball in R^n with centre x and radius ε.) So $F(P)$ does not have to be defined by such analytical formulas as (1.2), which can be useful, e.g., in problems discussed in Chapter 8. In the above formulation, we use sup instead of max as, in general, we do not know if the maximum of $f(x)$ is obtainable on $F(P)$, i.e., if $F^*(P) \neq \varnothing$.

The next generalization consists of considering the case in which only a part of the variables take integer values. We call such problems *mixed integer programming problems*. As an example of such a problem consider

$$v(P) = \max(cx + dy), \tag{P}$$

with restrictions

$$Ax + Dy \leqslant b, \quad x \geqslant 0, y \geqslant 0, x \in Z^n,$$

where A is the m by n matrix, D the m by p matrix, $d \in R^p$, $c \in R^n$, $b \in R^m$. If $n \geqslant 1$ and $p \geqslant 1$, then P is a *mixed integer programming problem*, but if $n = 0$, i.e., matrix A vanishes, then P is a *linear programming problem*. For $p = 0$, we have a *(pure) integer programming problem*.

Examples of such problems are given in Section 1.5 after introduction of basic equivalence relations between integer programming problems. In the bibliographical notes to this chapter, we give a short history of integer programming.

1.2. REFORMULATION OF INTEGER PROGRAMMING PROBLEMS

The idea of reformulation (transformation, reduction) of a given problem into an equivalent one is often used in mathematical programming since it is usually easier to solve or analyse such an equivalent problem by finding similarities and differences by comparison with the other problems. To introduce the idea of equivalence between integer programming problems, we consider the following

Example 1.1.

$$v(P) = \max(2x_1 + x_2) \tag{P}$$

sub   to

$$3x_1 + x_2 \leqslant 7,$$
$$2x_1 + 3x_2 \leqslant 10,$$
$$x \geqslant 0, \quad x \in Z^2.$$

Figure 1.1 gives a graphical interpretation of problem P. The feasible solutions are marked by circles and the objective function is pictured by a dashed line. One can see from the figure that infinitely many equivalent formulations of P exist, e.g., the

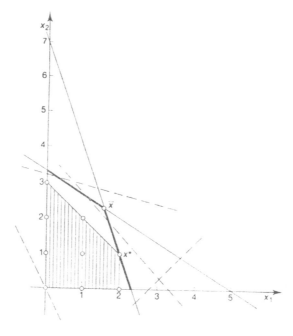

Fig. 1.1.

problem described by three constraints marked in Fig. 1.1 by dotted lines. In particular, $F(P)$ describes its convex hull, i.e., the smallest, in the sense of set inclusion, convex set containing $F(P)$, which we will denote as conv $F(P)$.

Set conv $F(P)$ is lined. Recall that S is convex if from $x_1, x_2 \in S$, $x_1 \neq x_2$, follows that a segment connecting x_1 with x_2 also belongs to S, i.e., $x = (\alpha x_1 + (1-\alpha)x_2)$ $\in S$ for any $0 \leqslant \alpha \leqslant 1$.

From Fig. 1.1 one can write constraints describing conv $F(P)$ and transform P into an equivalent problem P' of the form

$$v(P') = \max(2x_1 + x_2), \tag{P'}$$

subject to

$$x_1 + x_2 \leqslant 3,$$
$$x_1 \quad \leqslant 2,$$
$$x \geqslant 0, \quad x \in Z^2.$$

In other words, by solving P' we obtain the same result as solving P since $F(P')$ $= F(P)$ and $F^*(P') = F^*(P)$. Moreover, the constraint $x \in Z^2$ is redundant in P' and if we omit it, we obtain the linear programming problem P'' with $F^*(P'')$ $= F^*(P)$. $\qquad\qquad\qquad \square$

In the above transformations, only the constraints of P have been changed. We will discuss computational aspects of such transformations in Chapter 6.

We now give a few examples of equivalences which are true for any mathematical programming problem. Maximizing $f(x)$ with $x \in F(P)$ is equivalent to minimizing $g(x) = -f(x)$ with $x \in F(P)$. So we have

$$v(P) = \max_{x \in F(P)} f(x) = - \min_{x \in F(P)} (-f(x)).$$

Any constraint $g(x) \geqslant 0$ after multiplying by -1 is equivalent to $h(x) \leqslant 0$, with $h(x) = -g(x)$. An inequality $g(x) \leqslant 0$, after adding a nonnegative slack variable s, is transformed into an equivalent equality

$$g(x) + s = 0, \quad s \geqslant 0.$$

In integer programming problems, the integrality of s follows from the integrality of the data. An equality $g(x) = 0$ is equivalent to two inequalities $g(x) \geqslant 0$ and $g(x) \leqslant 0$. If x_j is not restricted in sign, then we may substitute it by the difference of two nonnegative variables, namely

$$x_j = x_j^+ - x_j^-, \quad x_j^+, x_j^- \geqslant 0.$$

In the case of a more general transformation, the objective function, constraints and variables may be changed. Before giving a definition of such an equivalence, we remind of the notion of an injective mapping (injection).

Every integer programming problem is defined in suitable chosen space, e.g. problem (1.1)–(1.4) is defined in R^n. Let X and Y be two spaces and let h be an injective mapping of set X on set Y, i.e., a mapping such that from $x_1 \neq x_2$ follows $h(x_1) \neq h(x_2)$ for any $x_1, x_2 \in X$. (We will denote such a mapping as $h: X \to Y$). Then there exists a reverse mapping $h^{-1}: Y \to X$ and it is an injection. By $h(A)$ we denote an image of set $A \subseteq X$ under the mapping h, i.e., $h(A) = \{y \in Y | y = h(x),$ $x \in A\}$.

DEFINITION 1.1. Problem P_1 defined in space X is equivalent to problem P_2 defined in space Y if there exists an injection $h: X \to Y$ such that $F^*(P_1) = h^{-1}(F^*(P_2))$ and $F(P_1) = h^{-1}(F(P_2))$.   □

We will now consider four examples of such transformations.

1.2.1. Transformation to Binary Problems

Binary problems, i.e., integer programming problems in which variables may take only two values, 0 or 1, play an important role in integer programming.

Below we show that every integer programming problem with bounded variables

$$v(P) = \max \{cx | Ax \leqslant b, 0 \leqslant x \leqslant d, x \in Z^n\} \tag{P}$$

may be transformed into the equivalent binary problem by substituting each integral variable by its binary representation. If $0 \leqslant x_j \leqslant d_j$, then

$$x_j = 2^0 y_1 + 2^1 y_2 + \ldots + 2^{r_j - 1} y_{r_j}, \quad y_j = 0 \text{ or } 1, \quad j = 1, \ldots, r_j,$$

where $2^{r_j - 1} \leqslant d_j \leqslant 2^{r_j} - 1$, $j = 1, \ldots, n$. Examples of such problems are given in Section 1.5.

The bounds $x_j \leqslant d_j, j = 1, \ldots, n$, are given in many problems or may be computed from the constraints $Ax \leqslant b$, as we show for the case of data from Example 1.1 in the following

Example 1.2. From the first constraint for $x_2 = 0$, we have $x_1 \leqslant 7/3$. Since $x \in Z^2$ and from the second constraint, we have $x_1 \leqslant 5$, then, finally, $0 \leqslant x_1 \leqslant 2$. In the same way we find that $0 \leqslant x_2 \leqslant 3$.

1.2.2. Linearization of the Nonlinear Problems

It follows from the above consideration that every bounded nonlinear integer programming problem can be transformed into an equivalent binary nonlinear integer programming problem. As for any p the pth power of a binary variable equals the variable, i.e., $x_j^p = x_j$, in a nonlinear binary problem we have only products of different variables.

Let

$$y_k = \prod_{j \in N_k} x_j = x_{j1} x_{j2} \ldots x_{jq} \tag{1.5}$$

be the kth product in such a problem, where N_k is the index set of this product and $q = |N_k|$, i.e., the cardinality of N_k equals q. The idea of a linearization method consists in substituting each product by a binary variable and adding two linear constraints which link the value of the binary variable with the value of the product. The constraints are of the form

$$\sum_{j \in N_k} x_j - y_k \leqslant |N_k| - 1, \tag{1.6}$$

$$\frac{1}{|N_k|} \sum_{j \in N_k} x_j - y_k \geqslant 0. \tag{1.7}$$

We show now that (1.6) and (1.7) determine exactly the same value of y_k as (1.5). If $x_j = 1$ for all $j \in N_k$, then from (1.6), $y_k \geqslant 1$ and (1.7) gives $y_k \leqslant 1$. So in this case, $y_k = 1$, exactly as in (1.5). If, on the other hand, at least for one $j \in N_k$, $x_k = 0$, then from (1.7), $y_k < 1$, and thus $y_k = 0$, again the same as in (1.5).

If in a binary nonlinear problem we have m constraints, n variables and r different products, then after such a linearization we have a binary linear problem with $m+2r$ constraints and $n+r$ variables.

1.2.3. *Constraint Aggregation*

We show now that it is possible to transform a bounded integer problem with an $m > 1$ equality constraints into an equivalent problem with only one equality constraint. Namely, we show that

$$v(P) = \max\{cx | Ax = b, 0 \leqslant x \leqslant d, x \in Z^n\} \tag{P}$$

is equivalent to the aggregated problem

$$v(P_a) \quad \max\{cx | uAx = ub, 0 \leqslant x \leqslant d, x \in Z^n\}, \tag{P_a}$$

i.e., we show that there exists an m-integer vector such that $F(P_a) = F(P)$. Such a vector will be called an *aggregation coefficients vector*.

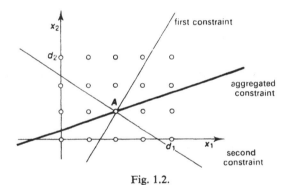

Fig. 1.2.

Figure 1.2 demonstrates this fact for the case $m = 2$, $n = 2$. There is only one feasible solution-point A on Fig. 1.2 in this case. The aggregated constraint (heavy line) passes through only one integer point for $0 \leqslant x \leqslant d$, namely, through point A. Moreover, the figure demonstrates that, in general, for a given system $Ax = b$, $0 \leqslant x \leqslant d$, there exist infinitely many equivalent aggregated constraints.

There are many methods of aggregation of integer equality constraints (diofantine equalities). For our purposes, we consider the method consisting of combining the first two equalities into an equivalent one and next aggregating it with the third and so on. Doing so $m-1$ times, we obtain problem P_a such that $F(P_a) = F(P)$.

Let

$$S \quad \{x \in Z^n \mid \sum_{j=1}^{n} a_{1j}x_j = b_1, \sum_{j=1}^{n} a_{2j}x_j = b_2, 0 \leqslant x \leqslant d\}.$$

Without loss of generality we may assume that $u_1 = 1$. We establish now conditions on u_2 such that set

$$S_a = \left\{ x \in Z^n \Big| \sum_{j=1}^{n} (a_{1j} + u_2 a_{2j}) x_j = b_1 + u_2 b_2, 0 \leqslant x \leqslant d \right\}$$

is equal to S. To determine u_2, we compute

$$h_1 = \max \left\{ \Big| \sum_{j=1}^{n} a_{1j} x - b_1 \Big| \Big| 0 \leqslant x \leqslant d, x \in Z^n \right\},$$

where $|x|$ is an absolute value of x. We may compute h_1 as a maximum of two numbers h_1^+ and $|h_1^-|$, i.e., $h_1 = \max\{h_1^+, |h_1^-|\}$, where

$$h_1^+ = \sum_{j=1}^{n} (\max\{0, a_{1j}\}) d_j - b_1,$$

$$h_1^- = \sum_{j=1}^{n} (\min\{0, a_{1j}\}) d_j - b_1.$$

We prove now

LEMMA 1.1. $S = S_a$ *if and only if u_2 is an integer such that $|u_2| > h_1$.*

Proof. We prove the lemma showing that $S \subseteq S_a$ (the necessary condition denoted as $\Rightarrow$) and $S_a \subseteq S$ (the sufficient condition, $\Leftarrow$).

($\Rightarrow$) If $x \in S$, then $x \in S_a$ for any u_2 not necessarily integer. Thus $S \subseteq S_a$.

($\Leftarrow$) Let $x' \in S_a$. Substituting x' into the second equality, we obtain

$$\sum_{j=1}^{n} a_{2j} x_j' = b_2 + k, \tag{1.8}$$

where k is an integer as all of the data in P are integer. Multiplying (1.8) by u_2 and subtracting it from

$$\sum_{j=1}^{n} (a_{1j} + u_2 a_{2j}) x_j' = b_1 + u_2 b_2,$$

we get

$$\sum_{j=1}^{n} a_{1j} x_j' = b_1 - k u_2. \tag{1.9}$$

From $x' \in S_a$ follows that $0 \leqslant x' \leqslant d$, and therefore

$$\Big| \sum_{j=1}^{n} a_{1j} x_j' - b_1 \Big| \leqslant h_1.$$

But $h_1 < |u_2|$, so in (1.9) must be $k = 0$, which by (1.8) means that x' satisfies (1.8) and (1.9). Since this is true for any $x' \in S_a$, then $S_a \subseteq S$ and the lemma is proved.   □

Now we can formulate

THEOREM 1.1. *Problem P is equivalent to problem P_a if there exist $u = (1, \beta, \beta^2, \dots$ $\dots, \beta^{m-1})$, where β is an integer such that $|\beta| > h$ with*

$$h = \max_i h_i = \max_i \left\{ \left| \sum_{j=1}^{n} a_{ij} x_j - b_i \right| \, \Big| \, 0 \leqslant x \leqslant d, x \in Z^n \right\}.$$

Proof. By repeating $m-1$ times the aggregation procedure given in the proof of Lemma 1.1, we find all of the aggregation coefficients. It is easy to see that the aggregation coefficients may grow as fast as an exponential function, so choosing β in such a way as in the theorem we get $F(P_a) = F(P)$. As the objective functions in both problems are the same, then both problems are equivalent. □

The exponential growth of aggregation coefficients as a function of m is the main drawback as it requires the storing of large numbers in a computer memory for which nonstandard ways of representation of such numbers are necessary. Although some aggregation methods may give smaller aggregation coefficients for a given particular problem, it can be proved (see Section 1.7 for references) that the aggregation coefficients must grow exponentially, in the worst case to reach $F(P_a)$ $= F(P)$.

Example 1.3. We aggregate the constraints from Example 1.1. After adding slack variables we get

$$v(P) = \max(2x_1 + x_2) \tag{P}$$

subject to

$$3x_1 + x_2 + x_3 \qquad = 7,$$
$$2x_1 + 3x_2 \qquad + x_4 = 10,$$
$$x \geqslant 0, \quad x \in Z^4.$$

In Example 1.2 we show that $0 \leqslant x_1 \leqslant 2, 0 \leqslant x_2 \leqslant 3$. In the same way, we get $0 \leqslant x_3 \leqslant 7, 0 \leqslant x_4 \leqslant 10$. Now we can compute $h_1^+ = 9, h_1^- = -7$. So if we put $u_2 = 10$, then

$$v(P) = \max(2x_1 + x_2),$$

under the constraint

$$23x_1 + 31x_2 + x_3 + 10x_4 = 107,$$
$$0 \leqslant x_1 \leqslant 2, \ 0 \leqslant x_2 \leqslant 3, \ 0 \leqslant x_3 \leqslant 7, \ 0 \leqslant x_4 \leqslant 10, \ x \in Z^4. \qquad □$$

1.2.4. *Optimization Problems in Finite Sets*

Consider an optimization problem in which each variable x_j, $j = 1, \dots, n$, takes a value from a finite set $D_j = \{k_{j1}, k_{j2}, \dots, k_{jn_j}\}$, where $n_j \geqslant 2$ is the cardinality of D_j. For each variable x_j, we introduce n_j binary variables $y_1, y_2, \ldots, y_{n_j}$ in the following way:

$$x_j = \sum_{i=1}^{n_j} k_{ji} y_i,$$

under a constraint

$$\sum_{i=1}^{n_j} y_i = 1.$$

So the problem is equivalent to a binary optimization problem.

1.3. BASIC RELAXATIONS OF INTEGER PROGRAMMING PROBLEMS

The notion of a relaxation of a given problem, similarly as the concept of equivalence, plays an important role in mathematical programming. The reason for its introduction is also the same — it simplifies analysis and comparison of a given problem with other problems, and quite often solving many relaxations of a given problem is easier than solving the problem in its given formulation (see Section 1.4).

DEFINITION 1.2. A problem

$$v(P_2) = \max_{x \in F(P_2)} g(x) \tag{P_2}$$

is a *relaxation* of a given problem

$$v(P_1) = \max_{x \in F(P_1)} f(x) \tag{P_1}$$

if

(i) $g(x) \geq f(x)$ for all $x \in F(P_1)$,
(ii) $F(P_2) \supseteq F(P_1)$.   □

For minimization problems, condition (i) takes the form $g(x) \leq f(x)$ for all $x \in F(P_1)$. In general, there are infinitely many relaxations of a given problem and we will study "good", in a certain sense, relaxations in Chapter 7. In this section, we introduce the most commonly used relaxations of integer programming problems.

More than thirty years of experience have proved that a linear programming problem obtained from a given integer programming problem after dropping the integrality requirements ($x \in Z^n$) is easier to solve than a given integer problem. Such a relaxation plays an important role in integer programming and therefore we denote it by a special symbol.

A linear programming problem

$$v(\bar{P}) = \max\{cx \mid Ax \leq b, x \geq 0\} \tag{$\bar{P}$}$$

is a *linear programming relaxation* (*LP-relaxation*) of a given integer programming problem

$$v(P) = \max\{cx \mid Ax \leq b, x \geq 0, x \in Z^n\}. \tag{P}$$

Thus, for a binary problem

$$v(P) = \max\{cx|Ax \leqslant b, x \in \{0, 1\}^n\}, \qquad (P)$$

the corresponding LP-relaxation is

$$v(\bar{P}) = \max\{cx|Ax \leqslant b, 0 \leqslant x \leqslant 1\}. \qquad (\bar{P})$$

By $\bar{x}$ we will denote an optimal solution to $\bar{P}$, thus $\bar{x} \in F^*(\bar{P})$. Obviously $\bar{P}$ is a relaxation of P as $F(P) \subseteq F(\bar{P})$ and the objective functions in both problems are the same.

The advantage of such a relaxation is the fact that if $\bar{x} \in Z^n$, then the solution of the relaxation is, at the same time, the solution of the given integer problem. Even if this is not the case, the value $c\bar{x}$ gives an upper bound on $v(P)$, namely

$$v(P) \leqslant \lfloor c\bar{x} \rfloor, \qquad (1.10)$$

where $\lfloor x \rfloor$ is an integer part of a real x, i.e., the largest integer not greater than x. In (1.10) we may take an integer part as all data in P are integers. Bound (1.10) is often used in integer programming methods.

Usually a problem with only one constraint is easier to solve than a problem with $m > 1$ constraints. For the system with $m > 1$ constraints

$$Ax \leqslant b,$$

the constraint

$$uAx \leqslant ub$$

is called a *surrogate constraint*, where $u \geqslant 0$ is a given vector in R^m. Then for a given integer programming problem

$$v(P) = \max\{cx|Ax \leqslant b, x \geqslant 0, x \in Z^n\} \qquad (P)$$

a *surrogate relaxation* (*problem*) takes the form

$$v(Su) = \max\{cx|uAx \leqslant ub, x \geqslant 0, x \in Z^n\}, \qquad (Su)$$

where again $u \geqslant 0$ is a given vector in R^m. It is easy to see that, by Definition 1.2, problem Su is a relaxation of P as $F(P) \subseteq F(Su)$.

We will also use a *Lagrangean relaxation* of a given problem which is constructed by putting all or part of the constraints into the objective function. More precisely, for a given problem

$$v(P) = \max\{cx|Ax \leqslant b, 0 \leqslant x \leqslant d, x \in Z^n\},$$

a Lagrangean relaxation is an integer programming problem of the form

$$v(Lu) = \max\{cx+u(b-Ax)|0 \leqslant x \leqslant d, x \in Z^n\}, \qquad (Lu)$$

where $u \geqslant 0$ is a given vector in R^m, called a vector of *Lagrangean multipliers*.

Obviously, we have $F(Lu) \supseteq F(P)$ and for all $x \in F(P)$ from $b-Ax \geqslant 0$ and $u \geqslant 0$ follows that $cx+u(b-Ax) \geqslant cx$ (condition (i) of Definition 1.2). Thus Lu is a relaxation of P. The feasible solution set of Lu is very simple and therefore Lu can be easily solved (see Chapter 7).

1.4. The Idea of Integer Programming Methods

If the feasible solution set of a given integer programming problem P is bounded, and this is true in any practical problem, then $F(P)$ contains a finite number of feasible points. Then one would suggest solving such a problem by an *explicit enumeration method*, which we describe for the case of binary problem

$$v(P) = \max\{cx|Ax \leqslant b, x \in \{0, 1\}^n\}. \qquad (P)$$

Starting from, say, $(0, 0, \ldots, 0)$ we check in turn whether a given binary vector x is feasible and, if it is, then we compare cx with the value of the best feasible solution found so far, called an *incumbent* and denoted as x^*. If $cx \geqslant cx^*$, then we set $x^* := x$ and $v(P) := cx^*$, where $:=$ means a substitution. At the beginning, we assume $v(P) := -\infty$. If after checking 2^n binary vectors we still have $v(P) = -\infty$, then P is inconsistent, otherwise x^* is an optimal solution.

Simple calculations show that even for $n = 100$, such a method requires 10^{19} days of work of a computer checking 10^6 binary vectors per second. Distribution of this work among all of the computers of the world does not help, and we must add that problems with thousands of variables are being solved today.

A natural development of the explicit enumeration method are *methods of implicit enumeration* or *branch-and-bound methods*. The idea of these methods consists in partitioning $F(P)$ into a number of subsets which is equivalent to dividing problem P into subproblems PP_i. Next we estimate $v(PP_i)$. Let x^* be the incumbent. Knowing these estimations and cx^* it is possible to say that some subproblems do not contain an optimal solution of P, i.e., to say that solving such subproblems up to optimality is not perspective. Such subproblems are removed from the list of subproblems. The perspective subproblems can be treated in the exactly the same way as problem P, i.e., each of them can be divided into some number of subproblems. Today, many methods of partitioning of $F(P)$ are known and the same is true for the estimation of $v(PP_i)$. We will study them in Chapter 4.

The second group of methods are *cutting-plane methods*. Their idea consists in solving a sequence of suitably constructed linear programming problems instead of solving a given integer programming problem. We explain it in Fig. 1.3, which is a repetition of Fig. 1.1. From Fig. 1.3 we see that the solution of the linear programming relaxation is not an integer as $\bar{x} = (11/7, 16/7)$. Therefore, we construct a cut (constraint) which cuts off $\bar{x}$ from $F(P)$ and add it to the set of constraints obtaining the integer programming problem P_1 having one more constraint than P. Next we solve $\bar{P}_1$ and since again $\bar{x} \notin Z^2$ we construct the second cut, which gives the problem P_2. From Fig. 1.3 we see that the solution of $\bar{P}_2$ is an integer and thus it is the solution of problem P.

In Chapter 3, we show that under mild assumptions a sequence of linear programming problems $\bar{P}, \bar{P}_1, \bar{P}_2, \ldots$ is finite. In a different methods, the cuts are constructed in a different way.

From Fig. 1.3 it follows that the rounding of $\bar{x}$ to the nearest integer is not an integer programming method since after rounding $\bar{x}$ we get vector (2.2), which is not even feasible in the problem from Example 1.1.

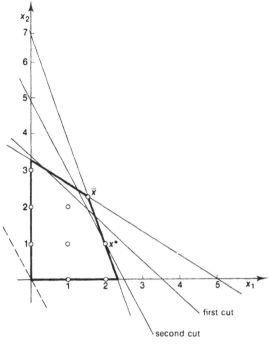

Fig. 1.3.

1.5. Applications of Integer Programming

So-called special problems, i.e., problems in which the objective function and/or constraints take a special form, play an important role in integer programming. Such problems often appear in practice; the strongest theoretical results have been obtained for them and special methods have been developed and analysed for them. The names of the problems reflect their applicability. We discuss below some such problems, giving first a verbal description and mathematical model for each problem and next consider some generalizations.

1.5.1. *The Knapsack Problem*

Consider a problem faced by a hiker packing n types (items) of equipment into his knapsack. Let $c_j > 0$ be the "value" and $a_j > 0$ be the "weight" of the jth item. If the total weight limitation (capacity of the knapsack) is b, then the problem of maximizing the total value of the packed knapsack may be formulated as the (binary) *knapsack problem*, abbreviated as K, namely

$$v(K) = \max \sum_{j=1}^{n} c_j x_j \qquad (K)$$

subject to

$$\sum_{j=1}^{n} a_j x_j \leqslant b,$$

$$x_j = 0 \text{ or } 1, j = 1, \dots, n,$$

where

$$x_j = \begin{cases} 1, & \text{if the } j\text{th item is packed,} \\ 0, & \text{otherwise.} \end{cases}$$

The problem has many applications in cargo loading, e.g., loading a spacecraft with scientific equipment, in project selection, and so on.

In a project selection problem, a_j is the cost of the jth project, while c_j is its profit and b is a total capital. In many practical problems, project selection is taken under many constraints of capital, energy, raw materials, etc. In such a problem, we have more than one constraint, so the problem has the form of (1.1)–(1.4), with all $a_{ij} \geqslant 0$ and $c_j \geqslant 0$. Such a problem is called a *multidimensional knapsack problem*, while the knapsack problem is sometimes called a *one-dimensional knapsack problem*.

If in the knapsack problem we substitute the constraint $x_j = 0$ or $1, j = 1, \dots; n$, by the constraint $x \geqslant 0$, $x \in Z^n$, then we obtain the *integer knapsack problem*. The problem constitutes the part of one important practical problem called the *cutting-stock problem*.

1.5.2. The Cutting-Stock Problem

The importance of the problem considered in this subsection is twofold: the cutting-stock problem is one of the first successful applications of integer programming in practice and its solution method shows how linear programming problems with a large number of variables can be efficiently solved.

For simplicity of the presentation, we consider here only the problem of one-dimensional cutting (for example rolls of paper) so that at least d_i pieces of length a_i are furnished, $i = 1, \dots, m$. The objective is to meet the demands, d_i, while minimizing the total number of rolls that must be cut. Since cutting each roll involves some waste, this keeps the total waste at a low level.

We assume that all rolls have the same length b such that $a_i \leqslant b$ for $i = 1, \dots, m$ and that we have an unlimited number of rolls. Then the problem is feasible. Define a cutting pattern as a particular way of cutting a length of stock, e.g. x_1 pieces of length a_1, x_2 pieces of length a_2, etc. There is a finite number of cutting patterns, although this number may be very large. For example, if $b = 200$ inches, $m = 40$ and $40 \leqslant a_i \leqslant 80$, $i = 1, \dots, m$, the number of patterns is over 10 million.

The problem may now be stated: Specify z_j, the number of times the jth cutting pattern should be used so that all demands are met and the total number of rolls used is minimized. If x_{ij} is the number of pieces of length a_i cut by the jth pattern and $p_j = (x_{1j}, \dots, x_{mj})$, $d = (d_1, \dots, d_m)$, then the *cutting-stock problem* may be formulated as

$$v(P) = \min \sum_j z_j \qquad\qquad (P)$$

subject to

$$\sum_j p_j z_j \geq d,$$

$z_j \geq 0$ and integer.

In many practical problems, the demands d_i are high and it is reasonable to expect that the optimal values of z_j are large enough so that a little is lost in rounding them to integers. So we may omit the integer requirements in P and solve P as a linear programming problem.

Such a problem is still difficult to solve due to the very large number of variables. The idea of the so-called *column generation method* is to start solving P with a small number of variables (columns) and add (generate) a new column if it is necessary. From linear programming theory (see Chapter 2) we know that in problem P a new, say, the jth, column is introduced into the basis if its relative cost coefficient

$$\bar{c}_j = 1 - \sum_{i=1}^{m} u_i x_{ij} \qquad (1.11)$$

is negative, where u_i are the simplex multipliers for the constraints of P.

In order that a vector $p_j = (x_{1j}, \dots, x_{mj})$ describes a legitimate cutting pattern, its coordinates must satisfy

$$\sum_{i=1}^{m} a_i x_{ij} \leq b, \qquad (1.12)$$

$x_{ij} \geq 0$ and integer. $\qquad (1.13)$

Thus the problem of minimizing $\bar{c}_j$ over all possible nonbasic columns becomes the integer knapsack problem

$$v = \max\left\{\sum_{i=1}^{m} u_i x_i \,\middle|\, \sum_{i=1}^{m} a_i x_i \leq b, x \geq 0, x \in Z^m\right\}. \qquad (1.14)$$

Observe that in formulation of this problem we may omit the index j in (1.11)–(1.13) and minimization of $\bar{c}_j$ in (5.13) is equivalent to maximization of

$$\sum_{i=1}^{m} u_i x_i.$$

If $v > 1$, then a new column $p = (x_1^*, \dots, x_m^*)$ is formed, where x_i^* is an optimal solution to (1.14), $i = 1, \dots, m$. This new column enters the basis and the linear programming problem with the number of variables increased by one is solved in the next iteration. It can be shown that after some number of iterations, $v \leq 1$ for all nonbasic columns, i.e., P is solved (see Chapter 2). So the integer knapsack problem plays an important role in solving the cutting-stock problem.

1.5.3. *The Capacitated Plant Location Problem*

The problem involves choosing locations for plants (facilities) producing a single product in such a way that the total cost of production and transportation of the product from the plants to the clients (consumers) in order to cover their demands are minimal.

Let $D_1, \ldots, D_m$ be plants with production capacity $a_1, \ldots, a_m$. Each D_i may produce a given product (may be opened, $x_i = 1$), or may not produce (may be closed, $x_i = 0$). Let c_i be a fixed cost of opening D_i, $i = 1, \ldots, m$, and d_{ij} be a unit cost of production in D_i and transportation from D_i to client C_j to cover his demand b_j, $j = 1, \ldots, n$. Now we formulate the *capacitated plant location problem* in the following way:

$$v(CL) = \min\left(\sum_{i=1}^{m} c_i x_i + \sum_{i=1}^{m}\sum_{j=1}^{n} d_{ij} y_{ij}\right) \tag{1.15}$$

subject to

$$\sum_{i=1}^{m} y_{ij} \geqslant b_j, \qquad j = 1, \ldots, n, \tag{1.16}$$

$$\sum_{j=1}^{n} y_{ij} \leqslant a_i x_i, \qquad i = 1, \ldots, m, \tag{1.17}$$

$$x_i = 0 \text{ or } 1, \qquad y_{ij} \geqslant 0, \qquad i = 1, \ldots, m, \qquad j = 1, \ldots, n. \tag{1.18}$$

The first sum in (1.15) corresponds to the total fixed cost, the second to the total production and transportation costs. By (1.16) the demand of each client is satisfied. The constraints (1.17) connect the values of the continuous variables y_{ij} with the values of the binary variables x_i, namely, if $x_i = 1$, then the amount of the product transported from D_i cannot be greater than a_i; if $x_i = 0$, then nothing is sent from D_i.

The problem is an example of a mixed integer problem. The case when a_i is sufficiently large, e.g.,

$$a_i > \sum_{j=1}^{n} b_i$$

is called the *simple (uncapacitated) plant location problem* (Exercise 1.8.7).

In Chapter 8, we will study problems with binary matrices of coefficients. One of them is the problem of (optimal job) assignment.

1.5.4. The Assignment Problem

If n jobs have to be done and if c_{ij} is the cost of training the jth worker to the ith job, then one may ask what is the cheapest assignment of workers to jobs.

Let $x_{ij} = 1(0)$ if the ith job is (is not) assigned to the jth worker. Now the *assignment problem* may be formulated as

$$v(AP) = \min \sum_{i=1}^{n}\sum_{j=1}^{n} c_{ij} x_{ij} \tag{AP}$$

subject to

$$\sum_{j=1}^{n} x_{ij} = 1, \qquad i = 1, \ldots, n,$$

$$\sum_{i=1}^{n} x_{ij} = 1, \qquad j = 1, \ldots, n,$$

$$x_{ij} = 0 \text{ or } 1, \qquad i = 1, \ldots, n, j = 1, \ldots, n.$$

Since all variables are binary, from the first constraints we have that each job is done by one worker, while the second group of constraints assures that each worker has one job.

The assignment problem often appears as a part (subproblem) of a more complicated problem as we show in the next subsection and has many generalizations. One of them is the *min-max assignment problem* which can be stated as

$$v(MMAP) = \min_{x \in F(AP)} \max_{i,j} c_{ij} x_{ij}. \qquad (MMAP)$$

This is an example of an integer programming problem with a nonlinear objective function. The feasible solution set of the problem is the same as in the assignment problem.

1.5.5. *The Travelling Salesman Problem*

Consider a problem faced by a travelling salesman who starting from city 1 wants to visit $n-1$ cities and come back to the starting point. In which order should he visit the cities to have his route as short as possible?

Let c_{ij} be the distance from city i to city j and let $x_{ij} = 1(0)$ if the travelling salesman goes (does not go) from i to j. In Chapter 8, we will show that, without loss of generality, we may require that each city is visited only once. Then the *travelling salesman problem* may be formulated as

$$v(TSP) = \min \sum_{i=1}^{n} \sum_{j=1}^{n} c_{ij} x_{ij} \qquad (TSP)$$

subject to

$$\sum_{j=1}^{n} x_{ij} = 1, \qquad i = 1, \ldots, n, \qquad (1.19)$$

$$\sum_{i=1}^{n} x_{ij} = 1, \qquad j = 1, \ldots, n, \qquad (1.20)$$

$$x_{ij} = 0 \text{ or } 1, \qquad i, j = 1, \ldots, n.$$

The above constraints and the objective function are the same as in the assignment problem, but their meaning is different. It follows from (1.19) and (1.20) that the travelling salesman leaves and visits respectively each city exactly once. A simple example with $n = 4$ and $x_{12} = x_{21} = 1$, $x_{34} = x_{43} = 1$, $x_{ij} = 0$ otherwise shows that these constraints are not sufficient. The graphical interpretation of this fact is given in Fig. 1.4. We see that instead of a tour, we have two subtours. We eliminate subtours if we require for any partition of the set of cities $N = \{1, \ldots, n\}$ into two

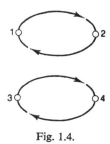

Fig. 1.4.

nonempty subsets the existence of at least one arc, i.e., $x_{ij} = 1$, connecting cities in different subsets. It can be written in the form

$$\sum_{i \in S} \sum_{j \in (N-S)} x_{ij} \geq 1 \tag{1.21}$$

for all $S \subseteq N$, $S \neq N$, $S \neq \emptyset$.

The travelling salesman problem has many applications in shop supply, post collecting, (water, electrical, etc.) installation design and in production scheduling. In Chapter 8, we will consider some generalizations of the problem.

1.5.6. The Quadratic Assignment Problem

Without loss of generality, we may assume that there is a set of n locations $D_1, \ldots, D_n$ in which n plants $W_1, \ldots, W_n$ must be constructed. The unit cost of shipping from location D_i to location D_k is d_{ik} and, in general, $d_{ik} \neq d_{ki}$. Also, the volume required to be shipped from plant W_j to plant W_l is q_{jl}. Let $x_{ij} = 1(0)$ if W_j is (is not) constructed on D_i. Then the *quadratic assignment problem* may be formulated as

$$v(QAP) = \min \sum_{i=1}^{n} \sum_{j=1}^{n} \sum_{k=1}^{n} \sum_{l=1}^{n} (d_{ik}q_{jl} + d_{ki}q_{lj}) x_{ij} x_{kl} \tag{QAP}$$

subject to the assignment constraints

$$\sum_{j=1}^{n} x_{ij} = 1, \qquad i = 1, \ldots, n,$$

$$\sum_{i=1}^{n} x_{ij} = 1, \qquad j = 1, \ldots, n,$$

$$x_{ij} = 0 \text{ or } 1, \quad i, j = 1, \ldots, n.$$

The problem has many applications in electrical wiring and design of keyboards, etc.

1.5.7. *Dichotomies and the Piecewise Linear Approximation*

Usually the feasible solution set of a given mathematical programming problem is the intersection of sets corresponding to the constraints, but in some applications, it is the sum of them.

Let a set $S_1 \subset R^n$ be described by the constraint $g(x) \leqslant 0$ and S_2 by $h(x) \leqslant 0$. Let $F(P) = S \cap (S_1 \cup S_2)$, where S is a bounded set in R^n. Then the problem

$$v(P) = \max_{x \in F(P)} f(x) \qquad (P)$$

may be written in the usual way if we substitute the condition

$$g(x) \leqslant 0 \text{ or } h(x) \leqslant 0 \text{ or both} \qquad (1.22)$$

by the system of constraints

$$g(x) \leqslant \delta \bar{g}, \qquad (1.23)$$
$$h(x) \leqslant (1-\delta)\bar{h}, \qquad (1.24)$$
$$\delta = 0 \text{ or } 1, \qquad (1.25)$$

where $\bar{g}$ and $\bar{h}$ are known finite upper bounds of $g(x)$ and $h(x)$ on S, respectively. To see this, we note that for $\delta = 0$, (1.23) gives $g(x) \leqslant 0$, while (1.24) gives the trivial constraint $h(x) \leqslant \bar{h}$. Similarly, for $\delta = 1$, we have $g(x) \leqslant \bar{g}$ and $h(x) \leqslant 0$. So (1.22) is equivalent to (1.23)–(1.25). We ask about a similar equivalence in Exercise 1.8.8.

Not only in mathematical programming do we often substitute a given nonlinear function by its piecewise linear approximation in a way as it was done with a single-valued function $f(x)$ defined on a segment (a, b) in Fig. 1.5. Let us assume that the breaking points are $z_1 = a, z_2, \ldots, z_p = b$. So we have

$$x = \sum_{k=1}^{p} \lambda_k z_k, \qquad (1.26)$$

$$g(x) = \sum_{k=1}^{p} \lambda_k f(z_k). \qquad (1.27)$$

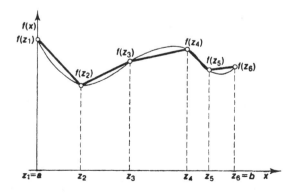

Fig. 1.5.

The function $g(x)$ is called the *piecewise linear approximation* of $f(x)$. The coefficients λ_k must satisfy the following conditions:

$$\sum_{k=1}^{p} \lambda_k = 1, \tag{1.28}$$

$$\lambda_k \geqslant 0, \quad k = 1, ..., p, \tag{1.29}$$

and moreover at most 2 of the λ_k are positive and these are of the form λ_k, λ_{k+1}. Similarly as in the beginning of this subsection, we introduce $p-1$ binary variables and $p+1$ constraints:

$$\lambda_1 \leqslant \delta_1, \tag{1.30}$$

$$\lambda_k \leqslant \delta_{k-1} + \delta_k, \quad k = 2, ..., p-1, \tag{1.31}$$

$$\lambda_p \leqslant \delta_{p-1}, \tag{1.32}$$

$$\sum_{k=1}^{p-1} \delta_k = 1, \tag{1.33}$$

$$\delta_k = 0 \text{ or } 1, \quad k = 1, ..., p-1. \tag{1.34}$$

If $1 \leqslant t \leqslant p-1$ and $\delta_t = 1$, then by (1.33) and (1.34), $\delta_k = 0$ for all $k \neq t$. Thus by (1.30)–(1.32), $\lambda_t \leqslant 1$, $\lambda_{t+1} \leqslant 1$ and $\lambda_k = 0$ otherwise.

1.6. A BRIEF INTRODUCTION TO COMPUTATIONAL COMPLEXITY

In the case when we have many methods, it is natural to ask which one is the best for a given problem or problems. This is a difficult question and in this section we present some results of computational complexity theory, which give at least a partial answer to it. The references for these results are given in Section 1.7. We consider now only exact methods, while near-optimal ones will be discussed in Chapter 9. One more remark: in this book we will distinguish between methods and algorithms, namely

algorithm = method + data structure,

although sometimes we will not specify the data structure assuming that it follows straight from a description of an algorithm.

1.6.1. *Complexity of Algorithms*

The *time-complexity* or, simply, *complexity* of an algorithm for solving some problem is said to be the maximal number of computational steps that it takes to solve any instance of the considered problem of a given size measured as the number of bits needed for encoding all data. As the maximum is taken over all problem instances, this is the *worst-case complexity*. The *expected-time complexity* is the average, overall input of a given size, of the number of computational steps required. As we will study only the worst case complexity, we will use the term complexity as an abbreviation of it.

Counting the number of computational steps is today, for the majority of known algorithms, an unrealistic task. Therefore, we study the complexity C_A of a given algorithm A as a function of the size s, so we study $C_A = f(s)$, more precisely, we study the *order* of $f(s)$, when $s \to \infty$.

Given two functions f and g whose domains are the natural numbers, we say that the order of f is lower or equal to the order of g, denoting this fact as $f = O(g)$ if

$$f(s) \leqslant kg(s)$$

for all $s \geqslant n_0$, where k and n_0 are positive constants. Then we briefly say that f is $O(g)$. Functions f and g are of the same order provided that $f = O(g)$ and that $g = O(f)$.

For instance, for the binary knapsack problem, it is reasonable to assume that its size s equals the number of variables n as the number of bits needed for encoding all data is proportional to n and perhaps to the right-hand side, b, but for simplicity of the analysis, we will for a moment assume that b is somehow bounded. Then the implicit enumeration algorithm has the complexity $O(2^n)$ as the number of operations, such as additions and comparisons, in the worst case is proportional to 2^n.

A *polynomial time algorithm* is defined as one with the complexity $O(p(s))$, where $p(s)$ is a polynomial function of s. In particular, a *linear time algorithm* has the complexity $O(s)$. An *exponential time algorithm* has a complexity which cannot be bounded by any polynomial of the size s. The difference between polynomial and exponential functions may be seen from the following example. Suppose an operation takes 1 μs, then for $n = 10$, the complexity n takes 10 μs, n^2 takes 100 μs and 2^n takes 1024 μs, while for $n = 60$, appropriate values are 60 μs, 3600 μs and 366 centuries!

1.6.2. *The Classes $\mathcal{P}$ and $\mathcal{NP}$*

Fortunately enough all polynomial time algorithms known so far have a complexity of a low order, e.g., n, $n \log n$, n^2, etc., not, say, n^{48}. So it is natural to distinguish between, generally speaking, easily solvable problems for which polynomial algorithms are known and difficultly solvable problems for which only exponential time algorithms are known. To make this more precise, we will consider decision problems which require a *yes* or *no* answer, instead of optimization problems. For instance, the knapsack optimization problem was formulated in Section 1.5 and we will write the corresponding decision problem in the following form:

instance: $n, b \in Z, a, c \in Z^n$
structure: There is $x \in \{0, 1\}^n$ such that $ax \leqslant b$ and $cx \geqslant k$, where $k \in Z$ is a given parameter. If $k \geqslant v(K) + 1$, then the answer to the above decision problem is *no*.

A problem type is in the *class $\mathcal{P}$* if there exists an algorithm that, for any instance, determines in polynomial time whether the answer is *yes* or *no*, i.e., its running time (the number of computational steps required) is bounded by a polynomial function of the problem size. A problem type is in the *class* $\mathcal{NP}$ if there exists an algorithm checking of a given structure in a polynomial time, thereby verifying the *yes* answer.

For example, finding the maximal number among n numbers requires $n-1$ comparisons, thus such a problem is in $\mathcal{P}$. Consider now the decision knapsack problem. If for a given x', $ax' \leqslant b$ and $cx' \geqslant k$, we verify the *yes* answer in $O(n)$ additions and comparisons. If it is not the case, i.e., if $ax' > b$ and/or $cx' < k$, then to verify the *no* answer, we have to check, in the worst case, all 2^n binary vectors.

Obviously, $\mathcal{P} \subseteq \mathcal{NP}$ and since many of the notorious combinatorial problems are among members of $\mathcal{NP}$ and are not known to belong to $\mathcal{P}$, it is commonly conjectured that $\mathcal{P} \neq \mathcal{NP}$.

The main result of the computational complexity theory establishes that certain combinatorial problems are of the same difficulty. To put it in a more precise way, we say that a problem P_1 is *reducible* to a problem P_2, and write $P_1 \propto P_2$, if P_1 can be considered as a special case of P_2, or more formally, if for any instance of P_1, a corresponding instance of P_2 can be constructed in polynomial time such that solving the later solves the former as well. A problem P_2 is said to be $\mathcal{NP}$-*complete* if $P_2 \in \mathcal{NP}$ and $P \propto P_2$ for all $P \in \mathcal{NP}$. Therefore, $\mathcal{NP}$-complete problems form a subclass of the most difficult problems in $\mathcal{NP}$ and moreover, they are *polynomially equivalent*, i.e., if P_2 is $\mathcal{NP}$-complete and $P_2 \propto P_3$ for some $P_3 \in \mathcal{NP}$, then P_3 is $\mathcal{NP}$-complete as well. If someone finds a polynomial time algorithm for a given $\mathcal{NP}$-complete problem, then such an algorithm would solve in polynomial time any problem in $\mathcal{NP}$ proving that $\mathcal{P} = \mathcal{NP}$, which is very unlikely.

If a decision problem is $\mathcal{NP}$-complete, then the corresponding optimization problem is said to be $\mathcal{NP}$-*hard* in the sense that it is at least as difficult as any problem in $\mathcal{NP}$.

The first problem showed to be $\mathcal{NP}$-complete is the satisfiability problem in which we ask if the set of boolean formulas has a truth assignment of boolean variables. It can be shown (Exercise 1.8.9) that this problem is reducible to the special case of the knapsack problem called the *partition problem* which can be formulated as

$$v(PP) = \max \sum_{j=1}^{n} a_j x_j \qquad\qquad (PP)$$

subject to

$$\sum_{j=1}^{n} a_j x_j \leqslant \frac{1}{2} \sum_{j=1}^{n} a_j,$$

$$x_j = 0 \text{ or } 1, \quad j = 1, \ldots, n.$$

We will conclude this section with a few remarks. It follows from Exercise 1.8.12 that for $\mathcal{NP}$-complete problems, an increment of the efficiency of computers by 10 or 100 times will only increase by a little the size of problem solved in, say, 1 hour of c.p.u. time. The history of integer programming shows that much better progress has been obtained by the development of new more efficient algorithms and new more efficient data structures. We will study this question in detail in Chapters 5 and 8. The same approach applied for problems from the class $\mathscr{P}$ gives in some cases algorithms whose complexities are equal to the complexities of the corresponding problems, i.e., they were the best possible algorithms.

We note that problems of quite similar formulations may belong to different complexity classes. For instance, the assignment problem belongs to class $\mathscr{P}$, while the travelling salesman problem, which differs from AP only in (1.21), belongs to class $\mathscr{N}\mathscr{P}$.

It is still unknown if some problems belong to class $\mathscr{P}$ or $\mathscr{N}\mathscr{P}$. In the next chapter, we will discuss a quite recent result (see Section 1.7) saying that the linear programming problem belongs to class $\mathscr{P}$.

1.7. BIBLIOGRAPHIC NOTES

The paper published by Gomory in 1958 is commonly considered as the beginning of discrete or integer programming. In 1960, Land and Doig published a description of the branch-and-bound method which became popular after application of it to the travelling salesman problem proposed by Little et al. (1963). Similarly, the Lagrangean relaxations became popular in integer programming after their application to the travelling salesman problem by Held and Karp in 1970. In 1972, Karp, basing on Cook's (1971) paper, introduced classes $\mathscr{P}$ and $\mathscr{N}\mathscr{P}$ and formulated the question whether $\mathscr{P} = \mathscr{N}\mathscr{P}$ (see also Karp, 1975). The second important question was the membership of linear programming in the class $\mathscr{P}$. A positive answer to this question was given by Khachian in 1979.

There are many books devoted to integer programming and among them the book by Garfinkel and Nemhauser (1972) is commonly considered as a basic one. The complexity of the integer and linear programming problem is studied in the book by Papadimitriou and Steiglitz (1982). The reviews by Garfinkel and Nemhauser (1973), Geoffrion and Marsten (1972), Balinski and Spielberg (1969) and Balinski (1965) are recommended. Geoffrion's (1974) paper can be also considered as a review.

The equivalence of integer programming problems is studied in Williams' (1978) book and his papers from 1974 and 1978 (see also Kaliszewski and Walukiewicz, 1981). The linearization method presented in Section 1.2 is from Watters (1967). The method for solving nonlinear integer problems was given by Lawler and Bell (1966), Hammer and Rudeanu (1968) and by Walukiewicz et al. (1973). A review of the aggregation methods is given by Rosenberg (1974). Babat (1979) proved that, in the worst case, the aggregation coefficients must grow not slower than an exponential function. Chvatal and Hammer (1977) proposed an aggregation method for some particular inequality constraints.

Garey and Johnson's (1979) book is commonly considered an excellent guide to computational complexity theory (see also a review paper by Slisenko (1981) and Lenstra et al. (1982)). Aho et al. (1974) is a good general introduction to design of efficient algorithms.

1.8. Exercises

1.8.1. n balls with weights $a_1, \ldots, a_n$ are put into two boxes in such a way that their total weights are as even as possible. Give at least three equivalent formulations of this problem.

1.8.2. Show that the travelling salesman problem is a particular case of the quadratic assignment problem.

1.8.3. Compute the aggregation coefficients for the assignment problem.

1.8.4. Consider the case of the cutting-stock problem when the number of rolls of paper is limited.

1.8.5. Formulate the so-called guillotine cutting problem in which rectangles are cut using guillotine cuts, i.e., cuts parallel to the sides of a rectangle.

1.8.6. In the simple plant location problem, put $y_{ij} \leqslant x_i$, $i = 1, \ldots, m$, $j = 1, \ldots, n$, instead of (1.17) and compare the linear programming relaxations corresponding to these two formulations of the problem.

1.8.7. Prove that the simple plant location problem has the single assignment property: there exists an optimal solution to the problem where each client receives all of his supply from a single facility only.

1.8.8. Find a system of constraints equivalent to the condition: if $g_1(x) \geqslant 0$, then $g_2(x) \geqslant 0$.

1.8.9. Show that the satisfiability problem is reducible to the partition problem.

1.8.10. Prove that the knapsack problem is $\mathcal{NP}$-hard.

1.8.11. Prove that the problem from Exercise 1.8.1 is $\mathcal{NP}$-hard.

1.8.12. Assume that we have four algorithms for solving a given problem with the size n. Let their complexities be $T_1 = 40n$, $T_2 = 2n\log n$, $T_3 = 0.5n^2$ and $T_4 = 2^n$ operations. Let a computer make 10^6 operations per second.
    (a) What will be the solution time of each algorithm for $n = 100$?
    (b) How big a problem may be solved by each algorithm within 1 hour of computation?
    (c) Let the speed of the computer be increased 10 times. What will be the answer to question (b)? How this answer changed if the speed increases 100 times?

CHAPTER 2

Linear Programming

In the first part of this chapter (Sections 2.1–2.7), we discuss the main results of linear programming which we will use in next chapters. The proofs of theorems are either sketched or left to be done in the exercises. They can be found in almost every textbook on linear programming. The second part (Sections 2.8–2.10) is devoted to the computational complexity of linear programming. In Section 2.8, we show that the simplex algorithm is an exponential, while in the next two sections we describe polynomial algorithms for linear programming problems.

2.1. BASIC SOLUTIONS

Without loss of generality, we can formulate the linear programming problem as

$$v(P) = \max \{cx | Ax = b, x \geq 0\}, \tag{P}$$

where A is a matrix with m rows and n columns, $c \in R^n$, $b \in R^m$. The feasible solution set of the problem, $F(P)$, as an intersection of a finite number of hyperplanes is a *convex polyhedron*. A point $x \in F(P)$ is an *extreme point* or *vertex of* a convex set $F(P)$ if there do not exist distinct $x_1, x_2 \in F(P)$ and a scalar α, $0 < \alpha < 1$, such that $x = \alpha x_1 + (1 - \alpha) x_2$.

As $F(P)$ is a closed, convex set and the objective function in P is continuous, then there are three mutually exclusive and collectively exhaustive possibilities:

(a) P is inconsistent, thus $F(P) = \emptyset$ and $F^*(P) = \emptyset$ too.

(b) There exists a vertex $\bar{x} \in F(P)$ and $y \in R^n$ such that $cy > 0$ and $(\bar{x} + \alpha y) \in F(P)$ for all $\alpha \geq 0$. Then $c(\bar{x} + \alpha y)$ and at the same time the problem P is unbounded and $F^*(P) = \emptyset$.

(c) There exists at least one vertex $\bar{x} \in F(P)$ such that

$$c\bar{x} \geq cx \quad \text{for all } x \in F(P),$$

so $\bar{x} \in F^*(P)$ and $v(P) = c\bar{x}$. If $\bar{x}_1, \bar{x}_2 \in F^*(P)$, then $x = \alpha \bar{x}_1 + (1 - \alpha) \bar{x}_2$ also is an optimal solution to P, providing $0 \leq \alpha \leq 1$.

Case (c) is the most important from a practical point of view, so we will start the analysis of P from it. Without loss of generality, we assume that $m \leq n$ and that the rows of A constitute the set of m linearly independent vectors, i.e., the equality

$$\sum_{i=1}^{m} \alpha_i a_i = 0$$

holds if and only if $\alpha_i = 0$, $i = 1, \ldots, m$, where a_i is the ith row of A.

We may always permute the columns of A in such a way that the first m columns of A form the set of linearly independent vectors. So we may write P in the form

$$v(P) = \max[c_B, c_N]\begin{bmatrix} x_B \\ x_N \end{bmatrix} \tag{P}$$

subject to

$$[B, N]\begin{bmatrix} x_B \\ x_N \end{bmatrix} = b,$$

$$x_B \geqslant 0, \quad x_N \geqslant 0.$$

In the above notation, B is called a *basic matrix* or, in short, a *basis*, while N is called a *nonbasic matrix*.

As the matrix B is nonsingular, the determinant of B, $\det B \neq 0$. The elements of $x_B = (x_{B1}, \ldots, x_{Bm})$ are called *basic variables*, while the elements of x_N are called *nonbasic variables*. By the same symbol B we will denote, in accordance with the linear programming tradition, a basis and the set of indices of vector x_B. If R is the set of indices of nonbasic variables, then $B \cup R = \{1, \ldots, n\}$, $|B| = m$ and $|R| = n - m$.

The constraints in P may be written as

$$Bx_B + Nx_N = b. \tag{2.1}$$

Since $\det B \neq 0$, then

$$x_B = B^{-1}b - B^{-1}Nx_N. \tag{2.2}$$

The particular solution of P

$$x_B = B^{-1}b, \quad x_N = 0 \tag{2.3}$$

is called a *basic solution* of P.

If $x_B \geqslant 0$, then (x_B, x_N) is called a *basic feasible solution* of P as $(x_B, x_N) \in F(P)$; the corresponding basis is called a *feasible basis*. A basic solution having at least one variable equal to zero is called *degenerate*.

It is known from linear algebra that a basic feasible solution corresponds to a vertex of the convex polyhedra $F(P)$. We note that many bases may correspond to a given vertex of $F(P)$ (Exercise 2.12.1). Two bases are called *adjacent* if they differ in only one column. As in P there are at most $\binom{n}{m}$ different bases, the linear programming problem may be solved in the following way: for a given basis we check if the corresponding solution is feasible, i.e., if $x_B \geqslant 0$. If this is the case, we compute $c_B x_B$ and update, if necessary, the value of the best feasible solution found so far and proceed to the next basis. If x_B has at least one negative coordinate, we proceed to the next basis.

The idea of the simplex method which we describe in the next section lies in the particular way of choosing the next basis, namely such that the basis satisfies two conditions:

(i) It is an *adjacent feasible basis*, i.e., it differs from a given feasible basis by only one column.

(ii) The value of the objective function for the adjacent basis is not worse (in our case not smaller) than the value for a given basis. We assume, obviously, that the first basis is feasible.

We summarize our consideration in the form of

THEOREM 2.1. $F(P) \neq \emptyset$ if and only if there exists at least one feasible basic solution. If P has an optimal solution, then it has a basic optimal solution. $\square$

2.2. THE SIMPLEX METHOD

We now describe the equivalent transformation of the problem P corresponding to the moving from a given basis to next adjacent one according to conditions (i) and (ii) from the previous section.

By x_{B0} we denote the value of the objective function at the point $x = (x_B, x_N)$ $\in F(P)$, where B is a given feasible basis of P. After substituting (2.2) into the objective function we have

$$x_{B0} = c_B B^{-1} b - (c_B B^{-1} N - c_N) x_N. \tag{2.4}$$

It is more convenient to write (2.2) and (2.4) in the matrix form

$$\begin{bmatrix} x_{B0} \\ x_B \end{bmatrix} = \begin{bmatrix} c_B B^{-1} b \\ B^{-1} b \end{bmatrix} - \begin{bmatrix} c_B B^{-1} N - c_N \\ B^{-1} N \end{bmatrix} x_N. \tag{2.5}$$

We denote the coordinates of vectors (2.5) in the following way:

$$h_0 = \begin{bmatrix} h_{00} \\ h_{10} \\ \vdots \\ h_{i0} \\ \vdots \\ h_{m0} \end{bmatrix} = \begin{bmatrix} c_B B^{-1} b \\ (B^{-1})_1 b \\ \vdots \\ (B^{-1})_i b \\ \vdots \\ (B^{-1})_m b \end{bmatrix}, \quad h_j = \begin{bmatrix} h_{0j} \\ h_{1j} \\ \vdots \\ h_{ij} \\ \vdots \\ h_{mj} \end{bmatrix} = \begin{bmatrix} c_B B^{-1} a_j - c_j \\ (B^{-1})_1 a_j \\ \vdots \\ (B^{-1})_i a_j \\ \vdots \\ (B^{-1})_m a_j \end{bmatrix}, \tag{2.6}$$

where $(B^{-1})_i$ denotes the ith row of the matrix B^{-1}, $i = 1, \ldots, m$, while a_j denotes the jth column of the matrix N.

Now we may write (2.5) in the form

$$x_{Bi} = h_{i0} - \sum_{j \in R} h_{ij} x_j, \quad i = 0, 1, \ldots, m, \tag{2.7}$$

where R is the set of indices of nonbasic variables.

Equations (2.7) express the value of basic variables in the terms of nonbasic variables. The zero row corresponds to the objective function, while the ith row corresponds to the ith constraint in P.

Assume now that we have a given basic feasible solution which is nondegenerate, so $x_{Bi} > 0$ for $i = 1, \ldots, m$. (We will consider the case of genenerate solutions in Section 2.4). We assume also that there exists at least one index $k \in R$ such that $h_{0k} < 0$. Introducing the kth column into the given basis, i.e., making $x_k > 0$, we increase by (2.7) the value x_{B0}, providing one column leaves the basis. In this way, we construct a new basis. As $x_j = 0$ for $j \in R - \{k\}$, we have

$$x_{Bi} = h_{i0} - h_{ik} x_k, \quad i = 1, \ldots, m, \tag{2.8}$$

and, if at least one $h_{ik} > 0$, then the new basis $\bar{B}$ is feasible ($x_{\bar{B}_i} \geq 0$) providing that

$$x_k \leq h_{i0}/h_{ik}.$$

This condition is satisfied if we put $x_k = \beta_{rk}$, where

$$\beta_{rk} = \frac{h_{r0}}{h_{rk}} = \min_{i=1,\ldots,m} \{h_{i0}/h_{ik} | h_{ik} > 0\}. \tag{2.9}$$

Then, by (2.8), $x_{Br} = 0$, which means that the rth column leaves the basis and the kth column enters the basis since $x_k = \beta_{rk} > 0$. The new basis is $\bar{B} = B \cup \{k\} - \{r\}$, which is adjacent to the given basis and it is feasible. For the new basis, we have

$$x_{\bar{B}0} = h_{00} - h_{0k} \beta_{rk} > h_{00} = x_{B0}.$$

So the value of the objective function increases.

The new basic solution is obtained by solving the rth equation (2.7)

$$x_k = \frac{h_{r0}}{h_{rk}} - \frac{x_{Br}}{h_{rk}} - \sum_{j \in R-\{k\}} \frac{h_{rj}}{h_{rk}} x_j, \tag{2.10}$$

and next, substituting (2.10) into (2.7) for $i \neq r$, we get

$$x_{\bar{B}i} = h_{i0} - \frac{h_{ik} h_{r0}}{h_{rk}} + \frac{h_{ik}}{h_{rk}} x_{Br} - \sum_{j \in R-\{k\}} \left(h_{ij} - \frac{h_{ik} h_{rj}}{h_{rk}} \right) x_j. \tag{2.11}$$

We get the new basic solution if we put $x_{Br} = 0$ and $x_j = 0$ for $j \in R - \{k\}$ in (2.10) and (2.11). One can easily check that it is feasible, i.e., that $x_{\bar{B}i} \geq 0$.

So far we assume that there exists in (2.9) at least one $h_{ik} > 0$. If $h_{ik} \leq 0$ for $i = 1, \ldots, m$, then, by increasing x_k, we will always have the feasible solution, and since $h_{0k} < 0$, then x_{B0} is unbounded and thus $F^*(P) = \emptyset$.

Let us consider the case when for a given basis B we have $h_{0j} \geq 0$ for all $j \in R$. Since $x_N \geq 0$ in P, so $c_B B^{-1} b$ is the upper bound for x_{B0} reached by the feasible solution $(B^{-1}b, 0)$. Thus $(B^{-1}b, 0) \in F^*(P)$, and $v(P) = cB^{-1}b$. Therefore we have proved

THEOREM 2.2. *Optimality Conditions. A basic solution* (2.7) *is an optimal solution to P if it satisfies two conditions*:

(i) $h_{0i} \not< 0$ for $i = 1, \ldots, m$ (*primal feasibility*),
(ii) $h_{0j} \geq 0$ for all $j \in R$ (*dual feasibility*).   □

We postpone the discussion of these conditions to Section 2.6.

Now we are ready to give a formal description of the simplex method (algorithm).

The Simplex Algorithm

Step 0 (Initialization): Find a starting basic feasible solution (x_B, x_N) with $x_B \geq 0$, $x_N = 0$ (see Section 2.3);

Step 1 (Optimality test): If $h_{0j} \geqslant 0$ for all $j \in R$, then $((x_B, x_N) \in F^*(P)$; $v(P)$ $:= c_B B^{-1} b$; stop), otherwise go to Step 2;

Step 2 (Choice of an entering variable): Select a variable x_k, $k \in R$, with $h_{0k} < 0$ to enter the basis; (A commonly used rule is $h_{0k} = \min\{h_{0j} | j \in R\}$);

Step 3 (Choice of a departing variable): If $h_{ik} \leqslant 0$ for $i = 1, \dots, m$, then stop (problem unbounded, $F^*(P) = \emptyset$), otherwise select x_r by (2.9) to leave the basis; (If there is a tie, break it arbitrarily, see Section 2.4);

Step 4 (Pivoting): Compute a new basic feasible solution by (2.10) and (2.11); $B := B \cup \{k\} - \{r\}$; $R := R \cup \{r\} - \{k\}$; go to Step 1. □

Each cycle between Steps 1 and 4 is called a *(simplex) iteration*. In the absence of degeneracy, we have $\beta_{rk} > 0$ in each iteration and therefore x_{B0} increases in each iteration. Thus we have

THEOREM 2.3. *If $F(P) \neq \emptyset$ and all basic feasible solutions are nondegenerate, then the simplex algorithm is finite as it finds an optimal solution in at most $\binom{n}{m}$ iterations.* □

2.2.1. *Simplex Tableaus*

It is useful to write all calculations of the simplex method in the form of the so-called *simplex tableaus*, e.g. Table 2.1.

TABLE 2.1

$$\dots -x_j \dots -x_k \dots$$

x_0	$h_{00} \dots h_{0j} \dots h_{0k} \dots$
$\vdots$	$\dots\dots\dots\dots\dots\dots\dots\dots$
x_{Bi}	$h_{i0} \dots h_{ij} \dots h_{ik} \dots$
$\vdots$	$\dots\dots\dots\dots\dots\dots\dots\dots$
x_{Br}	$h_{r0} \dots h_{rj} \dots h_{rk} \dots$
$\vdots$	$\dots\dots\dots\dots\dots\dots\dots\dots$

The zero column contains the basic feasible solution $x_{Bi} = h_{i0}$ for $i = 1, \dots, m$, while $x_0 = x_{B0} = h_{00}$ is the value of the objective function for this basic feasible solution. The zero row corresponds to the objective function of P and its elements h_{0j}, $j \in R$, are called the *simplex multipliers* (of a given basic feasible solution). The element h_{rk} playing a special role in (2.10) and (2.11) is called the *pivot (element)*. It is easy to verify that in the simplex algorithm all calculations are carried on in the tabular form using in Step 4 the following rules:

(1) Divide the rth row by the pivot element h_{rk}.

(2) Multiply the new rth row by h_{ik} and subtract it from the ith row, $i = 0, 1, \dots, m$, $i \neq r$.

(3) Replace the old kth column by its negative divided by h_{rk}, except for the pivot element which is replaced by $1/h_{rk}$.

As a result we obtain Table 2.2 from Table 2.1.

TABLE 2.2

		...	$-x_J$	...	$-x_{Br}$	...
x_0	$h_{00}-h_{0k}h_{r0}/h_{rk}$	...	$h_{0J}-h_{0k}h_{rJ}/h_{rk}$	...	$-h_{0k}/h_{rk}$	...
$\vdots$	....					
x_{Bi}	$h_{i0}-h_{ik}h_{r0}/h_{rk}$	...	$h_{iJ}-h_{ik}h_{rJ}/h_{rk}$	...	$-h_{ik}/h_{rk}$	...
$\vdots$	....					
x_k	h_{r0}/h_{rk}		... h_{rJ}/h_{rk}		... $1/h_{rk}$	...
$\vdots$	....					

Example 2.1. Consider the linear programming relaxation of the problem from Example 1.1

$$v(P) = \max x_0 = \max(2x_1 + x_2),$$
$$3x_1 + x_2 + x_3 \qquad = 7,$$
$$2x_1 + 3x_2 + \quad + x_4 = 10,$$

$x \geqslant 0$.

We add here two slack variables $x_3 \geqslant 0$, $x_4 \geqslant 0$ to have equalities. As $b > 0$, the slack variables give the feasible basis $B = \{3, 4\}$. We write the system in the form (2.7):

$$x_0 = 0 - 2(-x_1) - 1(-x_2),$$
$$x_3 = 7 + 3(-x_1) + 1(-x_2),$$
$$x_4 = 10 + 2(-x_1) + 3(-x_2),$$

and in the form of Table 2.3

TABLE 2.3

		$-x_1$	$-x_2$
x_0	0	-2	-1
x_3	7	3	1
x_4	10	2	3

We may make x_2 basic (Step 2) and x_4 nonbasic (Step 3). Thus we obtain the new basis $B = \{3, 2\}$ and Table 2.4, then Table 2.5.

TABLE 2.4

		$-x_1$	$-x_4$
x_0	10/3	$-4/3$	1/3
x_3	11/3	7/3	$-1/3$
x_2	10/3	2/3	1/3

TABLE 2.5

		$-x_3$	$-x_4$
x_0	38/7	4/7	1/7
x_1	11/7	3/7	$-1/7$
x_2	16/7	$-2/7$	3/7

Table 2.5 gives by Theorem 2.2 the optimal solution $x = (11/7, 16/7)$, $v(P) = x_0 = 38/7$. In Exercise 2.12.2, we ask about a graphical interpretation for this example.

2.3. THE FIRST BASIC FEASIBLE SOLUTION

Example 2.1 demonstrates how easily one can find the first basic feasible solution for the problem

$$v(P) = \{\max cx | Ax \leq b, x \geq 0\} \tag{P}$$

when $b \geq 0$. Suppose now that we have an equality constrained problem, $b \geq 0$ and A contains at least m unit vectors. Then the columns of A can be permuted to obtain the matrix (I, N) such that

$$x_B + N x_N = b,$$
$$x_B \geq 0, \quad x_N \geq 0,$$

where I is the m by m identity matrix. Obviously $x_B = b$, $x_N = 0$ is the first basic feasible solution.

Consider now the problem

$$v(P) = \max \{cx | Ax = b, x \geq 0\}, \tag{P}$$

in which A does not contain m unit vectors. Without loss of generality, we assume $b \geq 0$. We introduce the m-dimensional vector $x_s \geq 0$ of *artificial variables* and show that P has a basic feasible solution if and only if the problem

$$v(P_s) = \max \{cx | Ax + Ix_s = b, x \geq 0, x_s \geq 0\} \tag{P_s}$$

has a basic feasible solution with $x_s = 0$, i.e., when x_s is among nonbasic variables. Observe that for P_s the first basic feasible solution is $x_s = b$, $x = 0$.

The idea of the *two-phase method* consists in starting from $x_s = b$, $x = 0$, interchanging bases in such a way to arrive at the basic solution with $x_s = 0$ or proving that such a solution does not exist, which by Theorem 2.1 means that $F(P) = \emptyset$. This can be accomplished by solving

$$v(P') = \max(-1)x_s, \tag{P'}$$

subject to

$$x_0 - cx = 0,$$
$$Ax + Ix_s = b,$$
$$x \geq 0, \quad x_s \geq 0,$$

where 1 is the m-dimensional vector of ones. So in P' we minimize the sum of x_s. The vector $x_0 = 0$, $x_s = b$, $x = 0$ is the first basic feasible solution and we can solve P' by the simplex algorithm. Two cases are possible:

(1) $v(P') = 0$, which means that any optimal basis of P' is a first feasible basis for P as $x_s = 0$. The name of the method comes from the fact that in phase 1 we solve P' and in phase 2 we solve P. The artificial variables can be dropped as they leave the basis in phase 1, while the variable x_0 cannot be dropped in phase 1 as it is used in phase 2. Although the basic feasible solutions in phase 1 are degenerate since $x_0 = 0$, we can use the simplex method in the straightforward way, which we show in the next section.

(2) $v(P') < 0$, which means that $x_s \neq 0$, and then, by Theorem 2.1, $F(P) = \emptyset$.

2.4. Degenerate Solutions

Having a degenerate basic feasible solution with, e.g., $x_{Br} = 0$ and the kth column entering the basis by (2.9), we have $\beta_{rk} = 0$, and then the value of the objective function does not change with the change of basis since $x_{\bar{B}0} = h_{00} - h_{0k}\beta_{rk} = h_{00} = x_{B0}$. Thus the simplex algorithm may construct a sequence of bases $(B_1, B_2, ..., B_p)$ such that $B_1 = B_p$ and the value of the objective function is the same for all bases. So cycling in the simplex algorithm is possible and, in general, the simplex algorithm is not finite (Exercise 2.12.3).

Cycling occurs very rarely in practical problems and Exercise 2.12.3 demonstrates that very particular data are required to have such a phenomenon. In some proofs of integer programming theorems, though, we need a modification of the simplex algorithm which does not allow cycling. Such a modification is called the *lexicographic simplex method*.

A vector v is *lexicographically positive* (denoted as $v >_L 0$) if its first nonzero component is positive. A vector v is *lexicographically greater* than a vector u if $v - u >_L 0$. If $-v >_L 0$, then a vector v is *lexicographically negative*. The notation $v \geqslant_L 0$ means $v >_L 0$ or $v = 0$.

The lexicographic algorithm simplex differs from the simplex algorithm in the uniqueness of the choice of a variable leaving the basis. Due to the lexicographic order of vectors, the choice by (2.9) is unique. The lexicographic simplex uses the modified simplex tableau which is obtained from a given simplex tableau by inserting the mth order identity matrix between the zero column and the first column. For such a tableau, we have $v_0 = (h_{00}, 0, ..., 0)$, $v_i = (h_{i0}, 0, ..., 1, ..., 0)$ (the $(i+1)$st component equals 1), $i = 1, ..., m$. Obviously, these $(m+1)$-dimensional vectors are linearly independent and $v_i >_L 0$ for $i = 1, ..., m$.

Let $h_{0k} < 0$ and let

$$N_k = \{i | i \geqslant 1, h_{ik} > 0, h_{i0}/h_{ik} = \beta_{rk}\},$$

where β_{rk} is given by (2.9). Let $w_i = v_i/h_{ik}$, $i \in N_k$. If $|N_k| \geqslant 2$, then (2.9) is not unique and may produce a cycle of bases. We remove the tie by choosing x_{Br} in such a way that the corresponding vector w_r is lexicographically smallest among w_i, $i \in N_k$. Thus instead of (2.9) we have

$$w_r = \operatorname*{lex\,min}_{i \in N_k} w_i. \tag{2.12}$$

The choice in (2.12) is unique as all w_i, $i \in N_k$, are linearly independent. Thus the simplex algorithm with (2.12) instead of (2.9) is finite. We have to prove that the vectors $\bar{w}_i$, $i = 1, ..., m$, corresponding to the new basis obtained in Step 4 are lexicographically positive. We have $\bar{w}_r = w_r/h_{rk} >_L 0$ as $h_{rk} > 0$ and

$$\bar{w}_i = w_i - \frac{h_{ik}}{h_{rk}} w_r = w_i - h_{ik}\bar{w}_r \quad \text{for} \quad i = 1, ..., m, \ i \neq r.$$

If $h_{ik} \leqslant 0$, then $\bar{w}_i >_L 0$. If $h_{ik} > 0$, then, by (2.12),

$$w_i - w_r \geqslant \frac{v_i}{h_{ik}} - \bar{w}_r >_L 0 \quad \text{for} \quad i = 1, ..., m. \tag{2.13}$$

Multiplying (2.13) by $h_{ik} > 0$ we get $v_i - h_{ik}\bar{w}_r >_L 0$. Thus $\bar{w}_i >_L 0$ for $i = 1, ..., m$.

2.5. BOUNDED VARIABLES

Linear programming problems with bounded variables

$$v(P) = \max \{cx | Ax = b, p \leqslant x \leqslant d\} \qquad (P)$$

quite often arise in branch-and-bound methods. The bounds $x_j \geqslant p_j$ can be easily transformed to the standard form $x_j \geqslant 0$ by substitution of $x_j' = x_j - p_j$ for $j = 1, \ldots, n$. Therefore, we can consider the problem

$$v(P) = \max \{cx | Ax = b, 0 \leqslant x \leqslant d\}. \qquad (P)$$

If A has m linearly independent rows, then P may be solved by the simplex method by including $x \leqslant d$ into the constraints, which results in a problem with $m+n$ constraints and $2n$ variables. Such an approach is inefficient since computational experience indicates that the solution time depends on the number of constraints. Below we show that it is possible to modify the choice of the entering variable (Step 2) and the choice of the departing variable (Step 3) in such a way that P can be solved without the corresponding increase in the size of the basis.

A solution to $Ax = b$, $0 \leqslant x \leqslant d$ is called basic if each nonbasic variable x_j is equal to zero or d_j, $j \in R$. Let

$$R_1 = \{j \in R | x_j = 0\}, \, R_2 = \{j \in R | x_j = d_j\}.$$

So we have $R = R_1 \cup R_2$ and (2.7) takes the form

$$x_{Bi} = h_{0i} - \sum_{j \in R_1} h_{ij} x_j - \sum_{j \in R_2} h_{ij} x_j, \quad i = 0, 1, \ldots, m.$$

Since $x_j = d_j$ for $j \in R_2$, the basic solution is

$$x_{Bi} = h_{i0} - \sum_{j \in R_2} h_{ij} d_j = g_{i0}, \quad i = 1, \ldots, m \qquad (2.14)$$

and $x_j = 0$ for $j \in R_1$. This solution is feasible if

$$0 \leqslant g_{i0} \leqslant d_{Bi}, \quad i = 1, \ldots, m.$$

Furthermore, analogous to Theorem 2.2, we have

THEOREM 2.3. (*Optimality Conditions*). *A basic solution* (2.14) *is an optimal solution to P if*

(i) $0 \leqslant g_{i0} \leqslant d_{Bi}$ for $i = 1 \ldots, m,$
(ii) $h_{0j} \geqslant 0$ for all $j \in R_1,$
(iii) $h_{0j} \leqslant 0$ for all $j \in R_2.$ □

It follows from Theorem 2.3 that, in general, we get a better feasible basic solution if we introduce into the basis a nonbasic variable x_k such that if $k \in R_1$, then $h_{0k} \leqslant 0$, but if $k \in R_2$, then $h_{0k} \geqslant 0$, i.e., decreasing x_k from d_k results in increasing $x_{B0} = h_{00}$. A common rule to choose x_k is

$$h_{0k} = \max \{\max_{j \in R_2} h_{0j}, |\min_{j \in R_1} h_{0j}|\}. \qquad (2.15)$$

In the case of ties in (2.15), we may choose x_k arbitrarily. If in (2.15), e.g., $R_2 = \emptyset$, then we assume that

$$\max_{j \in R_2} h_{0j} = 0.$$

Therefore, we have to consider two cases: (1) $k \in R_1$, i.e., x_k entering the basis will increase from zero up to at most d_k, and (2) $k \in R_2$, i.e., x_k will decrease from d_k down to at most zero. The same cases should be considered for the variable departing from the basis.

Case 1: $k \in R_1$. By (2.14) we have

$$x_{Bi} = g_{i0} - h_{ik}\beta_k, \tag{2.16}$$

where β_k is an unknown increment of variable x_k. The increment should be satisfy the following constraints:

$$0 \leqslant g_{i0} - h_{ik}\beta_k \leqslant d_i, \quad i = 1, \ldots, m, \tag{2.17}$$

$$0 \leqslant \beta_k \leqslant d_k. \tag{2.18}$$

If for given $i = 1, \ldots, m$, $h_{ik} > 0$, then (2.17) reduces to $g_{i0} - h_{ik}\beta_k \geqslant 0$, which results in

$$\beta_k \leqslant \gamma_{rk} = \frac{g_{r0}}{h_{rk}} = \min_{i=1,\ldots,m} \{g_{i0}/h_{ik} | h_{ik} > 0\}. \tag{2.19}$$

Otherwise, if $h_{ik} < 0$, then (2.17) reduces to $g_{i0} - h_{ik}\beta_k \leqslant d_i$. Therefore,

$$\beta_k \leqslant \delta_{sk} = \frac{g_{s0} - d_s}{h_{sk}} = \min_{i=1,\ldots,m} \{(g_{i0} - d_i)/h_{ik} | h_{ik} < 0\}. \tag{2.20}$$

Combining (2.18), (2.19) and (2.20) we find that the increment of x_k from zero should be

$$\beta_k = \min \{\gamma_{rk}, \delta_{sk}, d_k\}. \tag{2.21}$$

If $\beta_k = \gamma_{rk}$ in (2.21), then x_{Br} is the departing variable. If $\beta_k = \delta_{sk}$, then x_{Bs} departs, while if $\beta_k = \delta_k$, then the basis does not change, but x_{B0} changes as the basic variable x_k grows from zero to d_k. If all $h_{ik} \leqslant 0$, then we put in (2.19) $\gamma_{rk} = +\infty$. Similarly, if all $h_{ik} \geqslant 0$, then $\delta_{sk} = +\infty$ in (2.20).

Case 2: $k \in R_2$. Now $\bar{\beta}_k$ is an unknown decrease of x_k from d_k to at most zero, i.e., $x_k = d_k + \bar{\beta}_k$, where $\bar{\beta}_k < 0$. The decrease should satisfy (2.16),

$$-d_k \leqslant \bar{\beta}_k \leqslant 0,$$

and if $h_{ik} < 0$, then from $g_{i0} - h_{ik}\bar{\beta}_k \geqslant 0$ we have

$$\bar{\beta}_k \geqslant \bar{\gamma}_{rk} = g_{r0}/h_{rk} = \max_{i=1,\ldots,m} \{g_{i0}/h_{ik} | h_{ik} < 0\}. \tag{2.22}$$

Otherwise, if $h_{ik} > 0$, then $\bar{\beta}_k$ has to satisfy constraints $g_{i0} - h_{ik}\bar{\beta}_k \leqslant d_i$. Therefore,

$$\bar{\beta}_k \geqslant \bar{\delta}_{sk} = (g_{0s} - d_s)/h_{sk} = \max_{i=1,\ldots,m} \{(g_{0i} - d_i)/h_{ik} | h_{ik} > 0\}. \tag{2.23}$$

Taking into account $\bar{\beta}_k \geqslant -d_k$ we finally get from (2.22) and (2.23)

$$\bar{\beta}_k = \max \{\bar{\gamma}_{rk}, \bar{\delta}_{sk}, -d_k\}. \tag{2.24}$$

Similarly as in (2.21), if $\bar{\beta}_k = \bar{\gamma}_{rk}$ in (2.24), then x_{Br} is eliminated from the basis; x_{Br} becomes a nonbasic variable with $x_{Br} = 0$. If $\bar{\beta}_k = \bar{\delta}_{sk}$ in (2.24), then x_{Bs} becomes a nonbasic variable with $x_{Bs} = d_s$. If $\bar{\beta}_k = -d_k$, then x_k remains a nonbasic variable, but now $k \in R_1$. Similarly as in Case 1, we put $\bar{\gamma}_{rk} = -\infty$ or $\bar{\delta}_{sk} = -\infty$ if in (2.22) or (2.23), respectively, the maximum is taken over the empty set.

Now we can modify the simplex algorithm for the case of bounded variables (Exercise 2.12.4).

2.6. DUALITY

For a given matrix A and vectors $b \in R^m$, $c \in R^n$, we may write two linear programming problems:

$$v(P) = \max \{cx | Ax \leqslant b, x \geqslant 0\}, \tag{P}$$

$$v(D) = \min \{ub | uA \geqslant c, u \geqslant 0\}. \tag{D}$$

Problem P is called the *primal problem*, while D is called the *dual problem*. In the above formulation, u is the m-dimensional row vector of *dual variables*, while x is a vector of *primal variables*.

If we have m inequalities in the primal problem, then in the corresponding dual one we have m nonnegative variables. The operator "max" changes for "min" in the dual problem. It is easy to show that the dual problem of D is the primal problem. Therefore, we may consider a primal-dual pair of problems of linear programming. Using the results of Section 1.2, one may write the dual problem for any form of a given linear programming problem (Exercises 2.12.5–2.12.7).

If $F(P) = \emptyset$, i.e., P is inconsistent, then we will assume that $v(P) = -\infty$. Similarly, if $F(D) = \emptyset$, then $v(D) = +\infty$.

THEOREM 2.4 (*Weak Duality*). $cx \leqslant ub$ for all $x \in F(P)$ and all $u \in F(D)$.

Proof. From $Ax \leqslant b$ and $u \geqslant 0$ we have $uAx \leqslant ub$. Similarly, from $uA \geqslant c$ and $x \geqslant 0$ we have $uAx \geqslant cx$. Thus $cx \leqslant ub$ for all $x \in F(P)$ and all $u \in F(D)$. □

THEOREM 2.5. *Exactly one of the following four cases holds*:
   (1) $F(P) = \emptyset$ *and* $F(D) \quad = \emptyset$,
   (2) $F(P) = \emptyset$ *and* $v(D) \quad = -\infty$,
   (3) $v(P) = +\infty$ *and* $F(D) = \emptyset$,
   (4) *there exist optimal solutions* $\bar{x} \in F^*(P)$ *and* $\bar{u} \in F^*(D)$ *and then* $v(P) = c\bar{x}$ $= v(D) = \bar{u}b$.

The fourth case is the most interesting and we formulate it in another way.

THEOREM 2.6. (*Strong Duality*). *If* $F(P) \neq \emptyset$ *and* $v(P) < +\infty$ *or if* $F(D) \neq \emptyset$ *and* $v(D) > -\infty$, *then there exists at least one optimal solution* $\bar{x} \in F^*(P)$ *and at least one optimal solution* $\bar{u} \in F^*(D)$ *and for each pair* $(\bar{x}, \bar{u})$ *we have* $\bar{u}b = c\bar{x}$. $\square$

Important conclusions concerning numerical methods and interpretation of the results of the methods can be drawn from the above theorems. For instance, by Theorem 2.4, we know that $cx \leqslant ub$ for any feasible solutions of P and D. By Theorem 2.6, the equality in the above inequality holds only for optimal solutions. Therefore, the optimal solutions satisfy the inequality $cx \geqslant ub$. So we have

COROLLARY 2.1. *A vector* $\bar{x}$ *is an optimal solution to P and a vector* $\bar{u}$ *is an optimal solution to D if and only if the vector* $(\bar{x}, \bar{u})$ *is a solution of the following system of linear inequalities in* R^{m+n}:

$$Ax \leqslant b,$$
$$-x \leqslant 0,$$
$$-uA \leqslant -c,$$
$$-u \leqslant 0,$$
$$ub - cx \leqslant 0. \qquad \square$$

The other relation between optimal solutions to P and D is given in

THEOREM 2.7. (*Complementarity Slackness*). *If* $\bar{x} \in F^*(P)$ *and* $\bar{u} \in F^*(D)$, *then*
  (i) $\bar{u}_i(b_i - a_i\bar{x}) = 0$, $i = 1, ..., m$,
  (ii) $\bar{x}_j(c_j - \bar{u}a_j) = 0$, $j = 1, ..., n$,
where $a_i(a_j)$ *is the ith row (the jth column) of A.* $\square$

We shall now give an economical interpretation of the above results. Consider the profit maximization problem from production of goods (commodities) from limited resources such as capital, energy, raw materials, etc. Let c_j be the unit production cost of the jth commodity, x_j the production volume of the jth commodity and a_{ij} the consumption of the ith resource for the production of a unit of the jth commodity, while b_i is the volume of the ith resource available. It follows from the equality $\bar{u}b = c\bar{x}$ that u_i is the unit price of the ith resource. Therefore, u_i, $i = 1, ..., m$, are often called the *dual prices* or *marginal prices*, or *shadow prices*.
  The dual prices have the following properties:
  (a) The dual prices are nonnegative, as $u \geqslant 0$ in D.
  (b) If for a given $\bar{x}$ the ith resource is not used ($a\bar{x} < b_i$), then the corresponding dual price is zero, which follows from Theorem 2.7 (i).
  (c) The value ua_j is the production cost of the jth commodity. The difference $c_j - ua_j$ may be considered as the unit profit. Since in D we have $c_j - ua_j \leqslant 0$, such a profit is always nonpositive.
  (d) The unit profit equals zero only for the optimal production vector $\bar{x}$ since by Theorem 2.7 (ii) we have $c_j - \bar{u}a_j = 0$ if the jth commodity is produced, i.e., when $\bar{x}_j > 0$.

The optimal values of the dual variables show how great the influence of the constraints on the objective function value at the optimal point is.

COROLLARY 2.2. *If $\bar{u}_i = 0$, the ith constraint is not tight (is redundant) at a given optimal point $\bar{x}$.* $\square$

The linear programming duality allows an optimal solution for D to be read from the optimal tableau for P. As P has m inequalities, then to solve it by the simplex method we have to put P in the form

$$v(P) = \max \sum_{j=1}^{n} c_j x_j \qquad (P)$$

subject to

$$\sum_{j=1}^{n} a_{ij} x_j + x_{n+i} = b_i, \quad i = 1, \ldots, m,$$

$$x_j \geqslant 0, \quad j = 1, \ldots, m+n.$$

The additional variables x_{n+i}, $i = 1, \ldots, m$, are sometimes called the *deficit variables*. The dual problem is

$$v(D) = \min \sum_{i=1}^{m} b_i u_i \qquad (D)$$

subject to

$$\sum_{i=1}^{m} a_{ij} u_i - u_{m+j} = c_j, \quad j = 1, \ldots, n,$$

$$u_i \geqslant 0, \quad i = 1, \ldots, m+n.$$

The additional variables u_{n+j}, $j = 1, \ldots, n$, are called the *surplus variables*. Thus in P and D we have $m+n$ variables and for a given constraint we have one decision variable and one additional variable. Thus we have *pairs of dually linked variables* x_j and u_{m+j} for $j = 1, \ldots, n$ and u_i and x_{n+i} for $i = 1, \ldots, m$.

Let $x = (x_1, \ldots, x_n)$, $x_d = (x_{n+1}, \ldots, x_{n+m})$ and let $u = (u_1, \ldots, u_m)$, $u_d = (u_{m+1}, \ldots, u_{m+n})$. Then the complementarity slackness conditions may be written in the form

$$\bar{u} x_d = 0, \qquad (i)$$

$$\bar{x} \bar{u}_d = 0, \qquad (ii)$$

or in the more developed form

$$\begin{cases} \bar{x}_j > 0 \Rightarrow \bar{u}_{m+j} = 0, \\ \bar{x}_{m+1} > 0 \Rightarrow \bar{u}_i = 0, \end{cases} \quad \text{and} \quad \begin{cases} \bar{u}_i > 0 \Rightarrow \bar{x}_{m+i} = 0, \\ \bar{u}_{m+j} > 0 \Rightarrow \bar{x}_j = 0. \end{cases} \qquad (2.25)$$

The reverse relations in (2.25) are, in general, not true, e.g., from $\bar{x}_j = 0$ does not follow $\bar{u}_{m+j} > 0$.

COROLLARY 2.3. *If $x_j(u_i)$ or $x_{n+i}(u_{m+j})$ is a basic variable in a given optimal solution, then in the optimal solution to the dual problem the dually linked variable $u_{m+j}(x_{n+i})$ or $u_i(x_j)$ equals zero.*   □

COROLLARY 2.4. *Having the optimal simplex tableau for P one can construct the optimal solution to D since the following hold:*

$$\bar{u}_i = \begin{cases} 0 & \textit{if } x_{n+i} \textit{ is a basic variable,} \\ h_{0,n+i} & \textit{if } x_{n+i} \textit{ is a nonbasic variable,} \end{cases} \quad i = 1, \dots, m,$$

$$\bar{u}_{m+j} = \begin{cases} 0 & \textit{if } x_j \textit{ is a basic variable,} \\ h_{0j} & \textit{if } x_j \textit{ is a nonbasic variable.} \end{cases} \quad j = 1, \dots, n. \qquad □$$

Example 2.2. The dual for the problem from Example 1.1 after introduction of the slack variable is

$$v(D) = \min(7u_1 + 10u_2) \tag{D}$$
$$3u_1 + 2u_2 - u_3 \quad = 2$$
$$u_1 + 3u_2 \quad - u_4 = 1$$
$$u \geqslant 0.$$

We may solve D by the simplex algorithm and check that $v(D) = v(P)$. We may also read the optimal solution to D from Table 2.5. By Corollary 2.4, we have

$$\bar{u}_1 = h_{0,2+1} = h_{0,3} = 4/7 \quad \text{as} \quad x_{2+1} = x_3 \text{ is a nonbasic variable,}$$
$$\bar{u}_2 = h_{0,2+2} = h_{0,4} = 1/7 \quad \text{as} \quad x_{2+2} = x_4 \text{ is a nonbasic variable,}$$

while $\bar{u}_3 = 0$ and $\bar{u}_4 = 0$ as both x_1 and x_2 are basic variables.   □

2.7. THE DUAL SIMPLEX ALGORITHM

We once more formulate the idea of the simplex algorithm: having a given primal feasible solution move to the next adjacent basis preserving the primal feasibility until the dual feasibility is reached (Corollary 2.4). By Theorem 2.2, the simplex tableau satisfying the conditions of primal and dual feasibility is the optimal one. One may use the reverse approach: having a dual feasible solution perform the simplex iterations until the primal feasibility is reached. This is the idea of the dual simplex algorithm.

Dual feasibility is easily obtainable in the cutting-plane method (see Chapter 3). Also in many practical problems we have $c \leqslant 0$, which can be used in construction of the first basic solution. For instance, consider the primal problem of the form

$$v(P) = \max\{cx | Ax - Ix_s = b, x, x_s \geqslant 0\}. \tag{P}$$

If at least one component of b is positive, no obvious primal basic feasible solution is available and we have to apply the two-phase method. However, if $c \leqslant 0$, there is an obvious dual feasible solution. Take $B = -I$ as the primal basis. Then the primal basic solution, not feasible, is $x_s = -b$ and $x = 0$. From $c_B = 0$ follows

$$h_{0j} = c_B B^{-1} a_j - c_j = -c_j \geqslant 0.$$

So we have a dual feasible solution and may apply the dual simplex algorithm.

Suppose we have a dual feasible solution, i.e., $h_{0j} \geqslant 0$ for $j \in R$, which is not primal feasible, and let $x_{Br} = h_{r0} < 0$. Then by dropping x_{Br} from the basis and introducing x_k, where k satisfies

$$\frac{h_{0k}}{h_{rk}} = \max_{j \in R} \left\{ \frac{h_{0j}}{h_{rj}} \Big| h_{rj} < 0 \right\}, \tag{2.26}$$

we obtain a new basis which is dual feasible, and in the absence of dual degeneracy ($h_{0j} > 0$ for all $j \in R$) the value of the dual objective function decreases.

If in the rth row $h_{rj} \geqslant 0$ for all $j \in R$, then the primal problem is infeasible and the dual problem is unbounded.

The Dual Simplex Algorithm

Step 0 (Initialization): Find the first dual feasible solution ($h_{0j} \geqslant 0$ for all $j \in R$);

Step 1 (Test of optimality): If $h_{i0} \geqslant 0$ for $i = 1, \ldots, m$, then stop ($\bar{x}_{B_i} = h_{i0}$; $\bar{x}_N = 0$); otherwise go to Step 2;

Step 2 (Choice of a departing variable): Select variable x_{Br} with $h_{r0} < 0$ to leave the basis; (A common rule is to choose x_{Br} is $h_{r0} = \min\{h_{i0}|i = 1, \ldots, m\}$);

Step 3 (Choice of an entering variable): By (2.26) select a variable x_k to enter the basis; If there is a tie, break it arbitrarily; If $h_{rj} > 0$ for all $j \in R$, then stop (D is unbounded and P is infeasible);

Step 4 (Pivoting): By (2.10) and (2.11), execute a dual simplex iteration; $B := B \cup \{k\} - \{r\}$; $R := R \cup \{r\} - \{k\}$; go to Step 1. $\qquad\qquad\square$

In the case of dual degeneracy, the lexicographic dual simplex algorithm can be constructed in a way similar as described in Section 2.4.

2.8. REMARKS ABOUT THE EFFICIENCY OF THE SIMPLEX METHOD

The question of the efficiency of the simplex algorithm was formulated at the very beginning of linear programming. More precisely, the question may be stated in the following way: what is the maximal number of simplex iterations needed to solve any linear programming problem with m constraints and n variables

$$v(P) = \max\{cx|Ax \leqslant b, x \geqslant 0\}. \tag{P}$$

If $I(P)$ denotes the number of the simplex iterations for P and if $I(P)$ depends on m and n, then we are asking about a function (estimation) $f(m, n)$ such that

$$I(P) \leqslant f(m, n) \tag{2.27}$$

for any data A, b, c such that the matrix A has m rows and n columns, $b \in R^m$ and $c \in R^n$. So we are interested in the worst-case analysis and indirectly in the computational complexity of linear programming (Section 1.6).

In this section, we will consider how the answer to the question has changed during the last 40 years. Due to the character of this section, we will give the references in the course of our presentation, not at the end of this chapter.

Since the number of bases is bounded by $\binom{n}{m}$, one may doubt the efficiency of the simplex method which considers bases in turn. Many years of computational experiments and thousands of solved problems form the ground for the following statement, sometimes quoted in linear programming textbooks: "on average, from $2m$ to $3m$ simplex iterations are needed to solve P". This is not the result of probabilistic analysis of the efficiency of the simplex algorithm which was undertaken recently (see e.g. Smale, 1983). If the above statement was true, the simplex algorithm would be a polynomial one.

In 1972, Klee and Minty constructed an example of a linear programming problem which requires an exponential number of simplex iterations to solve it (see also Clausen, 1980). The idea of the example is quite simple.

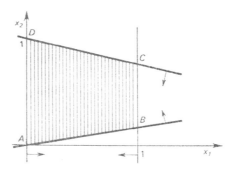

Fig. 2.1.

Let $F(P)$ be a "twisted" unit-box such as is given for case R^2 in Fig. 2.1. In a general case, $F(P)$ is described by $2n$ constraints, has 2^n vertices and may be written in the form

$$v(P) = \max x_n$$

subject to

$$
\begin{aligned}
-x_1 &\leqslant 0, \\
x_1 &\leqslant 1, \\
\varepsilon_1 x_1 - x_2 &\leqslant 0, \\
\varepsilon_1 x_1 + x_2 &\leqslant 1, \\
\cdots\cdots\cdots\cdots\cdots\cdots &\cdots\cdots\cdots\cdots \\
\varepsilon_{n-1} x_{n-1} - x_n &\leqslant 0, \\
\varepsilon_{n-1} x_{n-1} + x_n &\leqslant 1,
\end{aligned}
$$

where ε_i are given parameters such that $0 < \varepsilon_i < 1/2$ for each i.

If we solve P for $n = 2$ by the simplex algorithm, after one iteration the algorithm goes from A to D in Fig. 2.1, i.e., it finds the optimal solution. The construction of Klee and Minty transforms linearly $F(P)$ in such a way that the simplex algorithm goes through B, C and D, making $3 = 2^2 - 1$ iterations (Exercise 2.12.10). Therefore, the simplex algorithm is an exponential algorithm.

Charnes et al. (1980) noted that the dual problem to P can be easily solved. The dual takes the form

$$v(D) = \min(u_1^+ + u_2^+ + \ldots + u_n^+) \qquad (D)$$

subject to

$$-u_1^- + u_1^+ + \varepsilon_1 u_2^- + \varepsilon_1 u_2^+ \qquad\qquad\qquad = 0,$$
$$- u_2^- + u_2^+ + \varepsilon_2 u_3^- + \varepsilon_2 u_3^+ \qquad\qquad = 0,$$
$$\cdot\ \cdot\ \cdot\ \cdot\ \cdot\ \cdot\ \cdot\ \cdot\ \cdot\ \cdot\ \cdot\ \cdot\ \cdot\ \cdot\ \cdot\ \cdot\ \cdot\ \cdot\ \cdot$$
$$-u_{n-1}^- + u_{n-1}^+ + \varepsilon_{n-1} u_n^- + \varepsilon_{n-1} u_n^+ = 0,$$
$$-u_n^- + u_n^+ = 1,$$

$$u_i^-,\, u_i^+ \geqslant 0, \quad i = 1, \ldots, n.$$

It follows from the last constraint that $u_n^+ \geqslant 1$. Then the penultimate constraint gives $u_{n-1}^- > 0$ and next in turn we get $u_i^- > 0$ for $i = n-2, n-3, \ldots, 1$. The corresponding columns, together with the column corresponding to u_n^+, are linearly independent, and since there are n such columns, they form the basis. Moreover, from the minimization of the objective we have that $u_n^+ = 1$, $u_i^- = 1$ for $i = 1, \ldots, n-1$ is the optimal solution to D and $v(D) = 1$. From $\bar{x} = B^{-1}b$ we get the optimal solution to the primal $\bar{x}_n = 1$, $\bar{x}_i = 0$ for $i = 1, \ldots, n-1$. So we solve the dual problem in one simplex iteration. This example shows only that sometimes solving the dual is much easier than solving the primal problem. Obviously the example does not prove the polynomiality of the simplex algorithm for all possible linear programming problems.

The membership of linear programming in the class $\mathcal{NP}$ does not follow from the fact that the simplex method is exponential as other algorithms which solve the linear programming problem in polynomial time may exist. In fact, the possible membership of linear programming in the class $\mathcal{NP}$ was constantly doubted as this contradicts intuition supported by many years of computational experiments. Many authors conjectured that the linear programming problem belongs to $\mathcal{P}$. In 1979, Khachian finally resolved this problem by describing a polynomial algorithm for linear programming called the *ellipsoid algorithm*. We will study it in the next section.

Finally, we consider the numerical stability of the simplex method. In Step 4, we solve the system of m equalities with m unknowns $Bx_B = b$, and although $\det B \neq 0$, it is possible that $\det B$ is very close to zero. Numerical errors may then disturb result $x_B = B^{-1}b$, and we say that system $Bx_B = b$ is ill-conditioned.

The so-called *condition number* or the *measure of the conditioning of a matrix B* is defined in linear algebra as

$$\mathrm{cond}(B) = ||B||\,||B^{-1}||, \qquad (2.28)$$

where

$$||B|| = \max_{x \neq 0} ||Bx|| / ||x||$$

is the norm of a matrix B. It can be proved that $\mathrm{cond}(B) \geqslant 1$ for any matrix B and, in general, the larger the condition number, the more ill-conditioned matrix B is.

As far as a single system $Bx_B = b$ is concerned, we may compute (2.28) and solve an ill-conditioned system by a special method, but in linear programming we have to solve a sequence of linear systems and the above approach is unrealistic. Moreover, the numerical errors may, but do not have to, disturb the final result. In other words, if a given basis is ill-conditioned, the next basis may be well- or ill-conditioned. The above remarks show how difficult it is to define an appropriate condition number for a given linear programming problem. Finally, we note that changing the representation of data (e.g., writing each a_{ij} using many computer words instead of one) gives only partial help as there are problems (see Bednarczuk, 1977) for which such an approach is futile.

If a given basis is ill-conditioned, the constraints are almost parallel. A small perturbation may then produce a large change of coordinates of the corresponding vertex. Such almost parallel constraints appear in cutting plane methods discussed in Chapter 3.

2.9. THE ELLIPSOID ALGORITHM

In this section, we will consider the system of linear inequalities in R^n

$$(P) \qquad a_i^T x \leqslant \beta_i, \quad i = 1, \ldots, m, \tag{2.29}$$

where a^T denotes the transposition of a vector a.

By Corollary 2.1, we know that a given linear programming problem is equivalent to (2.29), which we will also denote by P. Without loss of generality, we may assume that in P:

(1) $a_i \neq 0$, $i = 1, \ldots, m$,
(2) $n \geqslant 2$,
(3) all data in P are integers.

If $a_i = 0$ and $\beta_i > 0$, then $a_i^T x \leqslant \beta_i$ is redundant. If $a_i = 0$, but $\beta < 0$, then $a_i^T x \leqslant \beta_i$ is inconsistent and $F(P) = \varnothing$, where $F(P)$ is the solution set of the system P. For $n = 1$, the system P may be, obviously, solved in polynomial time. The last assumption is necessary in the proof of polynomiality of the ellipsoid algorithm.

2.9.1. *The Idea of the Ellipsoid Algorithm*

Similarly as in the case of the simplex algorithm, we give a graphical interpretation of the ellipsoid algorithm. Consider the system of two inequalities denoted in Fig. 2.2 as (1) and (2).

We start by constructing an ellipsoid (a ball) E_0 centred at x_0 and with the radius ϱ choosen in such a way that E_0 contains at least one solution of P, if such exists. From a methodological point of view, we choose (1) and (2) in such a way that $F(P) \neq \varnothing$. Since $x_0 \notin F(P)$, in the next iteration we construct an ellipsoid E_1 in such a way that E_1 contains a part of E_0 in which may be a solution of P. As x_0 does not satisfy (2), the part of E_0 is lined in Fig. 2.2. The inequality (2) is called the *cut* as it cuts off $x_0 \notin F(P)$. Further, we show that the ellipsoid algorithm converges

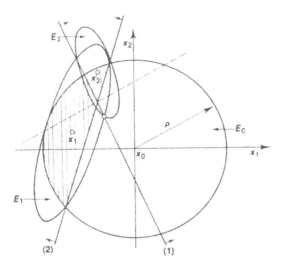

Fig. 2.2.

independently of the choice of the cut as, in general, there may be many constraints not satisfied by x_0, but the rate (speed) of the convergence heavily depends on the choice of the cut. Since x_1, the centre of E_1, does not satisfy (1), in the next iteration we construct an ellipsoid E_2 and check that its centre $x_2 \in F(P)$. Thus the ellipsoid algorithm finds a solution of P in this case in two iterations. From Fig. 2.2 one can see that the volumes of the constructed ellipsoids are smaller and smaller.

Let H_i be a half-space defined by the ith inequality, i.e., $H_i = \{x \in R^n | a_i^T x \leqslant \beta_i\}$. The ellipsoid E_k, $k = 0, 1, 2, \ldots$, is defined as

$$E_k = \{x \in R^n | (x - x_k)^T J_k^{-1} (x - x_k) \leqslant 1, \tag{2.30}$$

where x_k is the centre of E_k and J_k is a symmetric positive definite n by n matrix, i.e., $x^T J_k x \geqslant 0$ for any $x \in R^n$ and $x^T J_k x = 0$ if and only if $x = 0$. Now we define two principles for the construction of a sequence of ellipsoids $E_0, E_1, \ldots, E_k, \ldots$

The Inclusion Principle

$$\dot{E}_k \cap H_i \subseteq E_{k+1}, \quad k = 0, 1, \ldots,$$

where H_i corresponds to inequality $a_i^T x \leqslant \beta_i$ violated by x_k.

The Convergence Principle

$$\frac{\operatorname{Vol} E_{k+1}}{\operatorname{Vol} E_k} = r_k < 1, \quad k = 0, 1, \ldots,$$

where $\operatorname{Vol} E_k$ denotes the volume of E_k in R^n.

The above sequence of ellipsoids is unbounded, but we further show that from the integrality of data of P follows the finiteness of the sequence, i.e., after the construction of a finite number of ellipsoids either we find $x_k \in F(P)$ or conclude that $F(P) = \emptyset$.

In the next section, we show that basing on the above principles one may construct a sequence of balls, polytopes or, in general, any convex bodies which is convergent to a solution of P, if such exists. There we show that the sequence of ellipsoids has some advantages.

2.9.2. The Basic Iteration of the Ellipsoid Algorithm

It consists of moving from a given ellipsoid $E_k = (J_k, x_k)$ to the next ellipsoid $E_{k+1} = (J_{k+1}, x_{k+1})$, $k = 0, 1, \ldots$ To do this, we transform E_k into the unit ball described in coordinates $z_1, \ldots, z_n$ obtained from $x_1, \ldots, x_n$ by appropriate linear transformation.

It is known that any symmetric positive definite matrix J_k may be written in the form

$$J_k = Q_k^T Q_k, \tag{2.31}$$

where Q_k has real elements and $\det Q_k \neq 0$. Having a vector $a \neq 0$ we may transform E_k into the unit ball in the following way.

After the normalization of the vector $Q_k a$, we choose such an orthonormal matrix R_k ($R_k^T R_k = 1$) which transforms the vector into the unit vector e_1, i.e., into the first column of the unit matrix I. So we have

$$q_k = \frac{Q_k a}{||Q_k a||}, \qquad R_k q_k = e_1, \tag{2.32}$$

where $||a|| = \sqrt{a^T a}$ is the Euclidean norm of a.

We choose the linear transformation $h: R^n \to R^n$ as

$$z = R_k (Q_k^T)^{-1} (x - x_k). \tag{2.33}$$

After substituting of (2.33) into (2.30), by (2.31) we get $h^{-1}(E_k) = \{z \in R^n | z^T z \leqslant 1\}$. This transformation has the following geometrical interpretation. First, the operator R_k rotates the unit ball in such a way that q_k is transformed into e_1, and the next operator Q_k transforms the unit ball into the ellipsoid E_k with the centre in the origin of the system $x_1, \ldots, x_n$.

In the description of the ellipsoid algorithm, it is useful to define the *algebraic distance* of x_k from the half-space H_i

$$d_i(x_k) = \frac{a_k^T x_k - \beta_i}{||Q_k a_i||} = \frac{a_i^T x_k - \beta_i}{\sqrt{a_i^T J_k a_i}}. \tag{2.34}$$

One can easily prove the following

THEOREM 2.8. *One and only one of the following cases holds:*
  (i) *If $d_i(x_k) \leqslant -1$, then $H_i \cap E_k = E_k$ and the ith inequality is redundant in P.*
  (ii) *If $-1 < d_i(x_k) \leqslant 0$, then $H_i \cap E_k \neq \emptyset$ and x_k satisfies $a_i^T x \leqslant \beta_i$.*
  (iii) *If $0 < d_i(x_k) \leqslant 1$, then $H_i \cap E_k \neq \emptyset$ and x_k does not satisfy $a_i^T x \leqslant \beta_i$.*
  (iv) *If $d_i(x_k) > 1$, then $H_i \cap E_k = \emptyset$ and P is inconsistent.* □

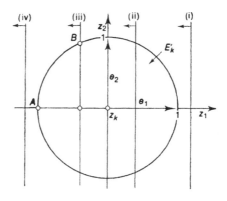

Fig. 2.3.

Figure 2.3 gives an interpretation of Theorem 2.8. Thus only in case (iii) do we have to construct the next ellipsoid $E_{k+1} = (J_{k+1}, x_{k+1})$. Let the new ellipsoid have the centre at point $z_{k+1} = -\gamma e_1$, where γ is a parameter the value of which we will specify later. Let the new ellipsoid have all but the first axes even. Then (2.30) takes the form

$$\frac{1}{\omega^2}(z_1 - z_{k+1,1})^2 + \frac{1}{\sigma^2}\sum_{j=2}^{n} z_j^2 \leq 1, \tag{2.35}$$

where ω, $\sigma > 0$ are parameters of this transformation and we will specify their value later. We write (2.35) in the matrix form

$$(z - z_{k+1})^T \begin{bmatrix} \dfrac{1}{\omega} & \\ & \dfrac{1}{\sigma}I_{n-1} \end{bmatrix} \begin{bmatrix} \dfrac{1}{\omega} & \\ & \dfrac{1}{\sigma}I_{n-1} \end{bmatrix}(z - z_{k+1}) \leq 1. \tag{2.36}$$

From (2.33) we have

$$z = (Q_k^T R_k^T)^{-1}(x - x_k) = R_k(Q_k^{-1})^T(x - x_k). \tag{2.37}$$

Substituting (2.37) into (2.36) and taking into account $z_{k+1} = -\gamma e_1$, we get

$$\frac{1}{\sigma^2}(x - x_{k+1})^T Q_k^{-1} R_k^T \begin{bmatrix} \dfrac{\sigma^2}{\omega^2} & \\ & I_{n-1} \end{bmatrix} R_k(Q_k^T)^{-1}(x - x_{k+1}) \leq 1, \tag{2.38}$$

where

$$x_{k+1} = x_k - \gamma Q_k^T \frac{Q_k a}{\|Q_k a\|} = x_k - \gamma \frac{J_k a}{\sqrt{a^T J_k a}}. \tag{2.39}$$

We want to write (2.38) in the form of (2.30). Since J_k is nonsingular, J_k^{-1} exists and one may check that

$$J_{k+1} = Q_{k+1}^T Q_{k+1} = \sigma^2 Q_k^T R_k^T \left(I - \left(1 - \frac{\omega^2}{\sigma^2} e_1 e_1^T \right) \right) R_k Q_k$$

$$= \sigma Q_k^T(I - \delta q_k q_k^T)\sigma Q_k = \sigma Q_k^T(I - \xi q_k q_k^T)^2 \sigma Q_k,$$

and finally we get

$$J_{k+1} = \sigma^2\left(J_k - \delta\,\frac{J_k a(J_k a)^T}{a^T J a}\right), \quad Q_{k+1} = \sigma\left(I - \xi\,\frac{Q_k a(Q_k a)^T}{\|Q_k a\|}\right)Q_k, \qquad (2.40)$$

where $\delta = 1-(\omega/\delta)^2$, $\xi = 1-(\omega/\sigma)$.

Let $a^T x \leqslant \beta$ be an inequality violated by x_k and the algebraic distance from x_k to $a^T x = \beta$ be $1 \geqslant \alpha > 0$. If we select $a^T x \leqslant \beta$ as a cut, from Fig. 2.3 follows that the ellipsoid of a minimal volume should pass by point A and therefore (2.35) takes the form

$$\frac{1}{\omega^2}(-1+\gamma)^2 = 1. \qquad (2.41)$$

The ellipsoid should also pass all points B of the form $(-\alpha, \pm\sqrt{1-\alpha^2}/(n-1),\ldots$
$\ldots, \pm\sqrt{1-\alpha^2}/(n-1)$. Then, by (2.35), we get

$$\frac{1}{\omega^2}(-\alpha+\gamma)^2 + \frac{1}{\sigma^2}(1-\alpha^2) = 1, \qquad (2.42)$$

which in turn gives

$$\omega = 1-\gamma, \quad \sigma = \sqrt{1+\alpha}\,\frac{1-\gamma}{\sqrt{1+\alpha-2\gamma}}. \qquad (2.43)$$

Now we can express the volume of E_{k+1} as a function of the parameter γ:

$$\mathrm{Vol}\,E_{k+1} = f(\gamma) = c\omega\sigma^{n-1} = c(1+\alpha)^{(n-1)/2}(1-\gamma)^n(1+\alpha-2\gamma)^{-(n-1)/2},$$

where c is a constant. To find γ^* for which the volume of E_{k+1} is the smallest possible, we solve the equality $\partial f/\partial\gamma = 0$ and get

$$\gamma^* = \frac{1+n\alpha}{1+n}. \qquad (2.44)$$

Substituting (2.44) into (2.43) we compute

$$\omega^* = \frac{n}{n+1}(1-\alpha), \quad \sigma^* = \sqrt{\frac{n^2}{n^2-1}(1-\alpha^2)} \qquad (2.45)$$

and

$$r_k(\alpha) = \frac{\mathrm{Vol}\,E_{k+1}}{\mathrm{Vol}\,E_k} = \left(\frac{n^2}{n^2-1}\right)^{(n-1)/2}(1-\alpha^2)^{(n-1)/2}\frac{n}{n+1}(1-\alpha). \qquad (2.46)$$

Therefore we have proved

THEOREM 2.9. *The ellipsoid of the minimal volume $E_{k+1} = (J_{k+1}, x_{k+1})$ constructed from a given ellipsoid $E_k = (J_k, x_k)$, $k = 0, 1, \ldots$, according to the inclusion principle and convergence principle is described by (2.39) and (2.40), where the optimal values of the parameters γ, ω, σ are given by (2.44) and (2.45).* $\square$

It follows from (2.46) that for $-1/n < \alpha \leqslant 1$ we have $r(\alpha) < 1$ and even for $-1 < \alpha < -1/n$, by (2.46), $r(\alpha) < 1$, but the smallest ellipsoid containing $\{x|a^T x \leqslant \beta\}\cap E_k$ is equal to E_k and then the convergence principle is violated.

For $\alpha = -1/n$, by (2.46), $\mathrm{Vol}\,E_{k+1} = \mathrm{Vol}\,E_k$. Therefore, the speed of convergence of the ellipsoidal algorithm is locally maximal if we choose as a cut·an inequality for which

$$\alpha = \max_i d_i(x_k). \tag{2.47}$$

The speed may be increased further if we construct E_{k+1} on a section of a ball as it is shown in Fig. 2.4. In Exercise 2.12.13, we ask about the formula for x_{k+1} and J_{k+1}.

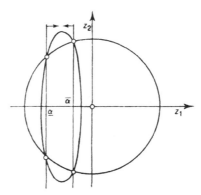

Fig. 2.4.

The next modification of the ellipsoid algorithm involves considering a surrogate constraint as a cut. Again we explain its idea in Fig. 2.2. We may consider a surrogate constraint for constraints (1) and (2) as a cut. It is denoted by a broken line in Fig. 2.2. It follows from Fig. 2.2 that $x_1 \in F(P)$, so the ellipsoid algorithm makes one iteration less in this case. It can be shown that finding the best surrogate constraint. i.e., constraint for which α in (2.47) is maximal, is equivalent to solving the quadratic programming problem (Exercise 2.12.14).

Solving the quadratic programming problem in each iteration may be too time consuming, therefore usually the surrogate constraint is computed in a near-optimal way, e.g., putting $u_i = 1$ or $u_i = \beta_i - a_i^T x_k$ for each constraint $a_i^T x \leqslant \beta_i$ violated at x_k, where u_i is a coefficient of the ith constraint in the surrogate constraint.

2.9.3. Polynomiality

Coding an integer p on a tape of a Turing machine or in the memory of a computer requires $1 + \lfloor \log p \rfloor$ bits if $p > 0$, and one bit if $p = 0$, where $\log p$ denotes the logarithm of the base 2 of p. Since all data in P are integers, we may count the length of data L, i.e., the number of symbols $+$, $-$, 0, 1, needed to code the data of P:

$$L = (2 + \lfloor \log m \rfloor) + (2 + \lfloor \log n \rfloor) + \left(2mn + \sum_{\substack{1 \leqslant i \leqslant m \\ 1 \leqslant j \leqslant n \\ a_{ij} \neq 0}} \lfloor \log |a_{ij}| \rfloor\right) +$$

$$+ \left(2m + \sum_{\substack{1 \leqslant i \leqslant m \\ b_i \neq 0}} \lfloor \log |b_i| \rfloor\right). \tag{2.48}$$

We assume that the data of P are coded in a specified order and the numbers are separated by symbols $+$ or $-$. In this subsection, we prove that the ellipsoid algorithm after at most $6n(n+1)L$ iterations finds a solution of P or shows that $F(P) = \emptyset$.

To simplify our considerations, we assume that all calculations are done in exact arithmetic. Since the number of iterations is bounded, one may compute the number of bits in the representation of a given number in such a way that rounding, if necessary, does not disturb the final result. For instance, it can be shown that it is sufficient to have $23L$ bits before the decimal point and $38nL$ bits after the decimal point.

The proof of polynomiality of the ellipsoid algorithm in fact consists of two observations.

(1) A solution of P, if it exists, is contained in the ball $S(0, 2^L)$, i.e., in the ball centred at the origin and with radius 2^L.

(2) If $F(P) \neq \emptyset$, then $F(P)$ contains the ball $S(a^*, 2^{-2L})$.

The second observation requires some comments. It is possible that a single point is the solution of P and then $\text{Vol}\, F(P) = 0$ in R^n. Therefore, we consider a system of strict inequalities

$$a_i^T x < \beta_i + \varepsilon, \quad i = 1, \ldots, m, \qquad (P')$$

where $\varepsilon = 2^{-L}$. In Lemma 2.1 we prove that having $x' \in F(P')$ one may construct in polynomial time $\hat{x} \in F(P)$. We also note that if $F(P) \neq \emptyset$, then $\text{Vol}\, F(P') > 0$, and, moreover, the ball $S(\hat{x}, 2^{-2L}) \subseteq F(P')$, where $\hat{x} \in F(P)$. Therefore, if we find that $E_k \subset S(a^*, 2^{-2L})$, then $F(P') = \emptyset$, and as a consequence $F(P) = \emptyset$. So we have to show that k is not growing faster than some polynomial of L.

In other words, we have to prove the following inclusions:

$$S(a^*, \tau) \subseteq P(P) \cap E_0 \subseteq S(0, \varrho).$$

From (2.46) for $k = 0, 1, \ldots$ we have

$$\frac{\text{Vol}\, E_{k+1}}{\text{Vol}\, E_k} < \frac{n}{n+1} \left(\frac{n^2}{n^2-1} \right)^{(n-1)/2} < e^{-(n+1)/2}.$$

Let us assume that the ellipsoid algorithm has made $k > 2n(n+1)\log(\varrho/\tau)$ iterations. Then the volume of E_k is smaller than $(\tau/\varrho)^n$ times the volume $S(0, \varrho)$, and it cannot contain the ball of the radius τ. From the inclusion principle we have

$$S(a^*, \tau) \subseteq F(P) \cap E_0 \subseteq E_k \quad \text{for} \quad k = 1, 2, \ldots \qquad (2.49)$$

This contradiction proves that if $S(a^*, \tau) \subseteq F(P) \cap E_0$, then $k \leqslant 2(n+1)\log(\varrho/\tau)$. So we have to prove that ϱ and τ are not growing faster than some polynomial of L.

Observation 1. First we prove that if $F(P) \neq \emptyset$, then $F(P) \cap S(0, 2^L) \neq \emptyset$, and therefore there exists $x \in F(P)$ such that $||x|| \leqslant 2^L$. After introduction of slack variables, we have the system of equalities $Ax = b$ and by Cramer's rule we get

$$|x_i| = \left| \frac{\det A_i}{\det A} \right| \leqslant \frac{|\det A_i|}{1}.$$

From Hadamard's inequality it follows that $\det A_i$ is not greater than the product of the norm of all columns A_i which may be estimated as

$$|x_i| \leqslant \frac{1}{mn} 2^L \leqslant \frac{1}{n} 2^L,$$

so $\|x\| \leqslant 2^L$ and $F(P) \cap E_0 \subseteq S(0, 2^L)$. Thus $\varrho = 2^L$.

Observation 2. Consider the first inclusion in (2.49).

LEMMA 2.1. *The system of linear inequalities*

$$a_i^T x \leqslant \beta_i, \quad i = 1, \ldots, m, \tag{P}$$

has a solution if and only if the system of strict inequalities

$$a_i^T x < \beta_i + \varepsilon, \quad i = 1, \ldots, m, \tag{P'}$$

has a solution, where $\varepsilon = 2^{-L}$.

Proof. If $x \in F(P)$, then $x \in F(P')$, so $F(P) \subseteq F(P')$. Let $x' \in F(P')$ and now we show how to construct $\hat{x} \in F(P)$ from x'. Let $M = \{i | \beta_i < a_i^T x' < \beta_i + \varepsilon\}$. We may assume that any column a_k is a linear combination of a_i, $i \in M$, i.e.,

$$a_k = \sum_{i \in M} p_{ki} a_i,$$

since otherwise, if a_k would be not a linear combination of a_i, $i \in M$, the system

$$a_i^T z = 0, \quad i \in M,$$
$$a_k z = 1$$

has a solution z'. Then taking $\bar{x} = x' + \lambda z'$ for sufficiently small λ we construct a new solution of P for which $|M|$ is greater by one. We can do this up to m times. Thus we may assume that for $k = 1, \ldots, m$

$$a_k = \sum_{i \in M'} p_{ki} a_i,$$

where $M' \subseteq M$ such that a_i, $i \in M'$, are linearly independent. By Cramer's rule

$$p_{ki} = D_{ki}/|D|,$$

where D_{ki} and D are determinants of appropriate matrices not greater than $(2^L)/mn$. Let $\hat{x}$ be a solution of the system

$$a_i^T x = \beta_i, \quad i \in M'.$$

Then for $k = 1, \ldots, m$, according to the definition of $\hat{x}$, we have

$$W = |D|(a_k^T \hat{x} - \beta_k) = \sum_{i \in M'} D_{ki} a_i^T \hat{x} - |D|\beta_k = \sum_{i \in M'} D_{ki} \beta_i - |D|\beta_k.$$

After adding and subtracting $|D| a_k^T x'$, we get

$$W = -\sum_{i \in M'} D_{ki}(a_i^T x' - \beta_i) + |D|(a_k^T x' - \beta_k) < \varepsilon \sum_{i \in M'} |D_{ki}| + \varepsilon|D|,$$

as $|a_i^T x' - \beta_i| < \varepsilon$ for $i \in M'$, and $|a_k^T x' - \beta_k| < \varepsilon$ for each k. Therefore, we get

$$W < 2^{-2L}(m+1)\frac{1}{mn}2^L < 1.$$

Since W is an integer, $W \leqslant 0$. So $a_k^T \hat{x} - \beta_k \leqslant 0$ for each k, i.e., $\hat{x} \in F(P)$.   □

If $x^* \in F(P')$, then $S(x^*, 1/\max\|2^L a_i\|)$ is contained in $F(P')$. As $\|a_i\| \leqslant 2^L$, we have $S(x^*, 2^{-2L}) \subseteq F(P')$ providing $F(P) \neq \emptyset$. Therefore

$$S(x^*, 2^{-2L}) \subseteq F(P') \cap S(0, 2^L) \subseteq S(0, 2^L).$$

So for P', we have $\tau = 2^{-2L}$, $\varrho = 2^L$, which gives

$$k \leqslant 2n(n+1)\log(\varrho/\tau) = 6n(n+1)L. \tag{2.50}$$

We note that the length of data of P' is at most equal to $(m(n+1)+1)L$, so it polynomially depends on L. From $F(P) = \emptyset$ if and only if $F(P') = 0$ follows that after $k = 6n(n+1)L$ iterations we may say that $F(P) = \emptyset$ if x_k does not satisfy all inequalities.

THEOREM 2.10. *The ellipsoidal algorithm requires at most $6n(n+1)L$ iterations to find a solution to P or to state that $F(P) = \emptyset$.*   □

COROLLARY 2.5. *The linear programming is in the class $\mathscr{P}$.*   □

Estimation (2.50) has no great practical value as even for small systems we obtain a very large number of iterations. Usually, the inconsistency of P may be detected earlier by Theorem 2.8 (see Section 2.11 for the references).

2.9.4. Solving the Linear Programming Problems

By Corollary 2.1, solving a given linear programming problem is equivalent to solving a system of linear inequalities. Such an approach has a few drawbacks. First, the volume of the set of solutions to P is equal to 0, as all solutions to P are on the hyperplane $cx = ub$. So we have to disturb the system in the appropriate way to apply the ellipsoid algorithm to it. Second, the number of inequalities is large, equal to $2(m+n)+1$. Third, if $F(P) = \emptyset$, we do not know whether the primal problem is unbounded or inconsistent. We have to add that, in general, the ellipsoid algorithm does not give the basic optimal solution. To construct such a solution, we may use Lemma 2.1 (Exercise 2.12.15). In two more efficient methods for solving linear programming problems of the form

$$v(PL) = \max\{c^T x | Ax \leqslant b, x \geqslant 0\}, \tag{PL}$$

we solve the parametric system of linear inequalities

$$\begin{aligned}
Ax &\leqslant b, \\
-x &\leqslant 0, \\
-cx &\leqslant -\xi,
\end{aligned} \tag{2.51}$$

where ξ is a given parameter. These methods do not give dual solutions.

Let us assume that by the ellipsoid algorithm we find x_k and $x_k \in F(PL)$. Then $\xi = c^T x_k$ is a lower bound on $v(PL)$. If $F(PL)$ is bounded and E_0 is choosen such that $F(PL) \subseteq E_0$, the upper bound $\bar{\xi}$ on $v(PL)$ may be estimated as

$$\bar{\xi} = c^T x_k + \sqrt{c^T J_k c}.$$

The idea of the *bisection method* consists in taking $\xi = (\xi + \bar{\xi})/2$ and in solving (2.51) for the new value of the parameter. There are two possible cases:

(i) The system (2.51) is consistent, i.e., x_{k+p} is a solution of (2.51) for $p \geqslant 1$. Then we put $\xi = c^T x_{k+p}$, compute $\bar{\xi}$ and again solve (2.51).

(ii) The system (2.51) is inconsistent. Then we go back to the ellipsoid E_k, put $\bar{\xi} = \xi$, compute ξ and solve (2.51). The computations may be stopped if ξ and $\bar{\xi}$ are sufficiently close. The comebacks to E_k are, obviously, the drawbacks of the method. The next method is free from them.

We assume again that $x_k \in F(PL)$ and $F(PL)$ is bounded. Then $\xi = c^T x_k$ and $-c^T x \leqslant -\xi$ is a cut with $\alpha = 0$ on which we can construct E_{k+1}. The idea of the *sliding objective function method* consists in making as big as possible step δ from x_k in the direction s in such a way that $x = x_k + \delta s$ is a feasible point for PL, where s is a direction of the fastest grow of $v(PL)$, i.e., $s = c$ or $s = J_k c$. The length of the step should satisfy

$$a_i x \leqslant \beta_i, \qquad i = 1, \ldots, m,$$
$$x = x_k + \delta s \geqslant 0.$$

The solution to the above system is

$$\delta = \min_i \frac{\beta_i - a_i^T x_k}{a_i^T s}, \qquad \delta \geqslant 0, \qquad x_k + \delta s \geqslant 0.$$

We may stop the computations if for a given $\varepsilon > 0$,

$$c^T \left(x_k + \frac{J_k c}{\sqrt{c^T J_k c}} \right) - c^T(x_k + \delta s) = \sqrt{c^T J_k c} - \delta c^T s \leqslant \varepsilon.$$

In both methods, from the inconsistency of $Ax \leqslant b$, $-x \leqslant 0$ follows $F(PL) = \emptyset$. The case $v(PL) = +\infty$ may be detected to showing the inconsistency of the dual problem.

In practical problems, the bounds $0 \leqslant x \leqslant d$ give a better estimation than $\varrho \leqslant 2^L$. Then we may put $x_0 = d/2$ and $\varrho = \|d\| \sqrt{n/2}$. Moreover, if $a_{ij} \geqslant 0$, $j = 1, \ldots, n$, from $a_i^T x \leqslant \beta_i$ we get

$$x_i^2 \leqslant \beta_i^2 / a_{ij}^2, \qquad j = 1, \ldots, n,$$

and therefore

$$\varrho \leqslant \sqrt{\sum_{j=1}^{n} \left(\frac{\beta_i}{a_{ij}} \right)^2}.$$

In many practical problems, we know in advance that $F(PL) \neq \emptyset$ and in consequence $F(P) \neq \emptyset$. Then we may choose ϱ in such a way that the largest algebraic distance $\alpha < 1$, for instance $\alpha = 1/2$. If in some iteration we obtain $\alpha > 1$, this does not mean $F(P) = \emptyset$ but only that the choice of ϱ was not good. Then we have to increase E_k to have again $\alpha < 1$. Computational experiments show the usefulness of such an approach.

2.10. SUBGRADIENT METHODS

In the previous section, we showed how to transform a given linear programming problem with inequality and/or equality constraints into a system of the form

$$a_i^T x \leqslant \beta_i + \varepsilon, \tag{2.52}$$

where ε is given. Now we give a general description of the ellipsoid algorithm.

Ellipsoid Algorithm
Step 0 (Initialization): Put $k := 0$ and choose x_0, ϱ according to Section 2.9;
Step 1 (Stop): If x_k satisfies (2.52), then stop;
Step 2 (Choice of the cut): By (2.47) choose the cut or construct a surrogate cut;
Step 3 (Iteration): By (2.39) and (2.40) compute x_{k+1}, J_{k+1} (or Q_{k+1}); put $k := k+1$ and go to Step 1.

Comparing the simplex algorithm with the ellipsoid algorithm we have to remember that we are comparing a particular method with a general one. The ellipsoid algorithm is one example of the so-called *subgradient methods* applied in nonlinear programming or nonlinear optimization. Consider a general nonlinear optimization problem of minimization without constraints of a function $f: R^n \to R$. In subgradient methods, an infinite sequence of points $x_0, x_1, \ldots$ in R^n is constructed such that

$$x_{k+1} = x_k + \gamma d, \tag{2.53}$$

where $d \in R^n$ is a direction of moving from x_k and γ is the length of the step in this direction.

Different subgradient methods differ in choice of γ and d and it may be proved that under weak assumptions the sequence of points (2.53) converges to the minimum of f if f is a convex function.

A function $f: R^n \to R$ is *convex* if the set of points of the form $(x, h) \in R^{n+1}$ with $h \geqslant f(x)$, called the *epigraph* of f, is convex (Exercise 2.12.16). If f is convex, then $g = -f$ is concave.

A convex function may be nondifferentiable. Therefore, the notion of the *subgradient* of a function f at a point x_k is introduced as a vector $g_k \in R^n$ such that

$$f(x) \geqslant f(x_k) + g_k^T (x - x_k). \tag{2.54}$$

The set of all subgradients at a given point x_k is called a *subdifferential* and denoted as

$$\partial f(x_k) = \{g_k \in R^n | f(x) \geqslant f(x_k) + g_k^T (x - x_k)\}.$$

One can easily check that finding at least one solution of the system

$$a_i^T x \leqslant \beta_i, \quad i = 1, \ldots, m,$$

is equivalent to minimization of the convex function

$$f(x) = \max_i(\beta_i - a_i^T x).$$

The ellipsoid algorithm is a subgradient method in which γ is choosen according to (2.44) and the direction d is computed as

$$d = \frac{J_k a}{\sqrt{a^T J_k a}},$$

where a is defined as the most violated inequality or the surrogate inequality.

The ellipsoid algorithm does not take into account that an optimal solution is in a vertex of the convex polyhedra. As we know, this observation is an essential one in the simplex method. The simplex algorithm is sensitive to the number of vertices of $F(P)$ and this number, as we show in Section 2.8, may grow exponentially with n. The simplex algorithm is also sensitive to the condition number of the basis matrix. The ellipsoid algorithm is free from these two drawbacks.

In the majority of the linear programing problems, the matrix A contains a few nonzero coefficients, usually not more than a few per cent of mm (such a matrix is called *sparse*), and often these nonzero coefficients form a special structure, e.g., a block-diagonal one. The simplex algorithm may be modified in such a way that the number of nonzero elements in each basis is more or less the same. This makes it possible to write a basis in the computer memory in an economical way and, in general, makes the simplex algorithm more efficient. Unfortunately, this is not the case with the ellipsoid algorithm. Even if A is a sparse matrix, then after just a few iterations J_k or Q_k becomes a dense matrix. This is a serious drawback of the ellipsoid algorithm that, up to date, has been removed only partially.

For simplicity, we write $J_k = \sigma^2 \bar{J}_k$. It is known that (2.40) may be written in the form

$$\bar{J}_{k+1} = \bar{J}_0 - \sigma_1 p_1 p_1^T - \sigma_2 p_2 p_2^T - \ldots - \sigma_k p_k p_k^T, \tag{2.55}$$

and similarly

$$Q_{k+1} = (I - \xi_1 q_1 q_1^T)(I - \xi_2 q_2 q_2^T) \ldots (I - \xi_k q_k q_k^T) Q_0. \tag{2.56}$$

So in each iteration we have to remember one scalar and one n-dimensional vector. Unfortunately, the ellipsoid algorithm usually requires more than n iterations and then (2.55) and (2.56) give no savings.

The so-called restart, the substitution of E_k by a ball, violates the convergence principle and decreases the speed of the convergence of the ellipsoid algorithm. On the other hand, remembering only the last l vectors in (2.55) or in (2.56) violates the inclusion principle and makes the ellipsoid algorithm only a near-optimal method.

Basing on the inclusion principle and the convergence principle one may construct a sequence of balls in a way similar as in the ellipsoid algorithm. Then, instead of (2.39), we have

$$x_{k+1} = x_k - \lambda \frac{a^T x_k - \beta}{a^T a} a. \tag{2.57}$$

The sequence x_k obtained by (2.57) is convergent for $0 < \lambda \leqslant 2$. For $\lambda = 2$, we obtain the *relaxation method*. It can be shown that the *ball method* is slower than the ellipsoid algorithm and moreover it is not a polynomial method (see Section 2.11).

In the ellipsoid algorithm we have

$$E_k \cap F(P) \subset E_{k+1}, \quad \text{but} \quad E_k \cap F(P) \neq E_{k+1}.$$

The obvious corollary from the above relations is used in the so-called *centred cross-sections method*.

Consider minimization of a convex function f defined on a convex polytope P_0 in R^n. In the centred cross-sections method, we construct a sequence of points $x_0, x_1, \ldots$ and a sequence of convex polytopes $P_0, P_1, \ldots$, where x_k is the gravity centre of P_k and P_{k+1} is defined as

$$P_{k+1} = \{x \in P_k | g_k^T x \leqslant g_k^T x_k\},$$

with g_k being a subgradient of f at x_k. Since f is convex, P_{k+1} contains all points $x \in P_k$ for which $f(x) \leqslant f(x_k)$. As x_k is the gravity centre of P_k, it can be proved (see Section 2.11 for references) that

$$\frac{\text{Vol} P_{k+1}}{\text{Vol} P_k} \leqslant \left(1 - \frac{1}{e}\right) < 1,$$

thus the convergence principle is satisfied. Moreover, P_{k+1} contains exactly all candidates for the minimum of f so in this method

$$P_k \cap F(P) = P_{k+1} \quad \text{for all } k.$$

Unfortunately, computing the gravity centre of P_k in R^n for $n \geqslant 3$ is a difficult task. As new cuts $g_k^T x \leqslant g_k^T x_k$ are added, the number of vertices of P_k is, in general, growing very fast, which additionally complicates the calculation of the gravity centre. So we may conclude that the ellipsoid algorithm combines the relatively high convergence rate with the simplicity of computations in each iteration.

2.11. BIBLIOGRAPHIC NOTES

2.1–2.8. Dantzig's book published in 1963 is commonly considered a classic on linear programming. We also suggest books by Gass (1964), Chvatal (1983) and Schrijver (1986).

Different criteria for the choice of variables introduced into the basis are studied by Goldfarb and Reid (1977). Bland (1977) describes a modification of a simplex algorithm which excludes cycling (see also Kotiah and Steinberg, 1978).

2.9. Khachian published in 1979 a description of the ellipsoid algorithm and proved that linear programming is in the class $\mathscr{P}$. This development has sparked enormous interest and Wolfe (1980), for instance, gives about 40 references written on that subject within almost a year. Bland et al. (1981) give a good review, while Gill et al. (1981) and Waluk and Walukiewicz (1981) published results of computational experiments with the ellipsoid algorithm. In 1984, Karmarkar published a poly-nomial time projection algorithm for linear programming which seems to have some practical value.

2.10. The subgradient methods are described, e.g., in the book by Shor (1985). Motzkin and Schoenberg (1954) describe the relaxation method, while Goffin (1979) studies the efficiency of the method of balls. Levin (1965) gives a description of the centred cross-sections method. It is interesting to note that Yudin and Nemirovskii in 1976 proposed the ellipsoidal algorithm as a method removing some drawbacks of the centred cross-section method.

2.12. EXERCISES

2.12.1. Show that many bases may correspond to a given vertex.

2.12.2. Construct the dual to the problem from Example 2.1 and solve it by the simplex algorithm. Give a graphical interpretation of the work of the simplex algorithm in both cases. In Example 2.1, choose x_1 as the entering basis in the first iteration and complete the calculations.

2.12.3. Check that we have cycling in the problem

$$v(P) = \max(150x_1 - 75x_2 + x_3 - 3x_4)$$
$$25x_1 - 6000x_2 - 4x_3 + 900x_4 + 100x_5 = 0,$$
$$25x_1 - 4500x_2 - x_3 + 150x_4 + 50x_6 = 0,$$
$$x_3 + x_7 = 1,$$

$$x_j \geqslant 0, \quad j = 1, ..., 7.$$

2.12.4. Write the simplex algorithm for the case of bounded variables.

2.12.5. Formulate the dual problem to

$$v(P) = \max \sum_{i=1}^{n} \sum_{j=1}^{m} c_{ij} x_{ij},$$

$$\sum_{i=1}^{m} x_{ij} = b_j, \quad j = 1, ..., n,$$

$$\sum_{j=1}^{n} x_{ij} = a_i, \quad i = 1, ..., m,$$

$$x_{ij} \geqslant 0, \quad i = 1, ..., m, j = 1, ..., n.$$

2.12.6. Construct the dual problem to the linear programming relaxation of the simple plant location problem.

2.12.7. Write the dual to

$$v(P) = \max \sum_{k=1}^{m} \sum_{j \in N_k} c_j x_j,$$

$$\sum_{k=1}^{m} \sum_{j \in N_k} a_j x_j \leqslant b,$$

$$\sum_{j \in N_k} x_j = 1, \quad k = 1, \ldots, m,$$

$$x_j \geqslant 0, \quad j \in N_k, k = 1, \ldots, m.$$

2.12.8. Prove Theorem 2.6.

2.12.9. Prove Theorem 2.7.

2.12.10. Construct a linear programming problem showing nonpolynomiality of the simplex method.

2.12.11. Solve the problem from Exercise 2.12.10 for $n = 3$.

2.12.12. Prove Theorem 2.8.

2.12.13. Find formulas for the ellipsoid constructed on a segment of a ball.

2.12.14. Find the surrogate constraint with the maximal algebraic distance from a given point (Goldfarb and Todd, 1980).

2.12.15. Having $x' \in F(P)$ (Section 2.9) construct a basic solution $x \in F(P)$ such that $\|x - x'\| \leqslant \varepsilon$, where ε is given (Bland et al., 1981).

2.12.16. For a convex function $f : R^n \to R$ defined on a closed set X, show that
  (a) f attains its minimum at some $x^* \in X$;
  (b) $\partial f(x)$ is nonempty and convex for any $x \in X$;
  (c) the zero vector $0 \in \partial f(x^*)$.

CHAPTER 3

Unimodularity and Network Flows. Cutting-Plane Methods

At the beginning of this chapter, we show that the assumption of the integrality of data (Section 1.1) has a deep theoretical meaning. Next, in Sections 3.2–3.4 we consider problems in which requirements of integrality are redundant. By solving these problems by the simplex method we find an integer optimal solution. If integrality requirements are not redundant, by adding appropriate cuts we may transform solving of a given integer programming problem into solving a sequence of linear programming problems. The remainder of this chapter is devoted to the description of cutting-plane methods.

3.1. INTEGRALITY OF DATA

In Section 2.1, we have shown that for a linear programming problem

$$v(\bar{P}) = \max\{cx|Ax \leqslant b, x \geqslant 0\}, \tag{$\bar{P}$}$$

exactly one of the following three possibilities must hold:

(a) $F(\bar{P}) = \emptyset$, i.e., $\bar{P}$ is infeasible,

(b) $v(\bar{P}) = \infty$, i.e., $\bar{P}$ is unbounded, $F^*(\bar{P}) = \emptyset$,

(c) $\bar{P}$ has an optimal solution $\bar{x} \in F^*(\bar{P})$, i.e., $cx \leqslant c\bar{x}$ for all $x \in F(\bar{P})$.

The example below demonstrates that for the integer programming problem

$$v(P') = \max\{cx|Ax \leqslant b, x \geqslant 0, x \in Z^n\}, \tag{P'}$$

without the assumption about the integrality of data, the above trichotomy, in general, does not hold.

Example 3.1. Consider the following integer problem:

$$v(P') = \max(-\sqrt{2}x_1 + x_2) \tag{P'}$$

subject to

$$-\sqrt{2}x_1 + x_2 \leqslant 0,$$
$$x_1 \geqslant 1,$$
$$x_2 \geqslant 0,$$
$$x_1, x_2 \in Z.$$

A portion of the feasible region for P' is shown in Fig. 3.1. P' is not infeasible as, e.g., $(1, 0) \in F(P')$. It is neither unbounded since it follows from the first constraint

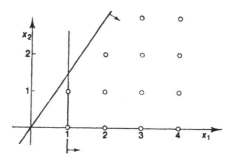

Fig. 3.1.

that $v(P') \leqslant 0$. Thus neither (a) nor (b) hold and we now show that (c) does not hold. To do this, it is sufficient to show the existence of a sequence of feasible points $(x_1^{(k)}, x_2^{(k)})$, $k = 1, 2, \ldots$, such that

$$0 \leqslant \sqrt{2} - \frac{x_2^{(k)}}{x_1^{(k)}} < \frac{1}{(x_1^{(k)})^2}, \quad x_1^{(k)} \to \infty$$

(Exercise 3.9.1). Then

$$\sup_{k \to \infty} (-\sqrt{2} x_1^{(k)} + x_2^{(k)}) = 0.$$

But no integer point satisfies the equality

$$-\sqrt{2} x_1 + x_2 = 0.$$

Therefore neither (a) nor (b), nor (c) hold for problem P'.   □

Consider now a general mixed integer problem

$$v(P) = \max(cx + dy), \qquad\qquad (P)$$

with the equality constraints

$$Ax + Dy = b,$$

$$x, y \geqslant 0, \quad x \in Z^n,$$

where A is an m by n matrix, D is an m by p matrix, $c \in R^n$, $d \in R^p$, $b \in R^m$. Below we formulate sufficient conditions for the above trichotomy.

THEOREM 3.1. *If at least one of the following three conditions is fulfilled*:
  (1) $p = 0$, *i.e., P is a (pure) integer problem*,
  (2) *all coefficients in A and in D are rational numbers*,
  (3) *there exists a constant M such that if* $(x, y) \in F(P)$, *then* $\|x\| \leqslant M$,
P must satisfy exactly one of the conditions (a)–(c).   □

Below we sketch the proof of Theorem 3.1 (see Section 3.8 for references). If condition (1) holds, either $F(P) \neq \varnothing$ or $F(P) = \varnothing$. In the last case, (a) holds. If $F(P) \neq \varnothing$, then either $v(P) < \infty$ or $v(P) = \infty$ and, in this case, (b) holds. Thus, we have to consider the case $F(P) \neq \emptyset$ and $v(P) < \infty$. It can be proved that if condition (1) of Theorem 3.1 is satisfied, $F(P)$ contains a finite number, say r, of extreme points. Let

$$v^* = \max_{1 \leqslant k \leqslant r} cx^{(k)},$$

where $x^{(k)}$ is the kth extreme point of $F(P)$. Then every $x \in P$ may be written as

$$x = \sum_{k=1}^{r} \alpha_k x^{(k)} + q,$$

where $q \in H = \{x | Ax = 0, x \geqslant 0, x \text{ rational}\}$ and

$$0 \leqslant \alpha_k \leqslant 1, \quad \sum_{k=1}^{r} \alpha_k = 1.$$

We show now that if x^* is an extreme point of $F(P)$ such that $cx^* = v^*$, then $x^* \in F^*(P)$. If $x \in F(P)$, then, by $v(P) < \infty$, we have $cq \leqslant 0$ and

$$cx \leqslant c \sum_{k=1}^{r} \alpha_k x^{(k)} = \sum_{k=1}^{r} \alpha_k (cx^{(k)}) \leqslant \sum_{k=1}^{r} \alpha_k v^* = v^*.$$

So, if $cx^* = v^*$, then $x^* \in F^*(P)$ and (c) holds.

If condition 2 is satisfied, it can be proved that $\text{conv} F(P)$ is polyhedral, i.e., the intersection of a finite number of half-spaces (Exercise 3.9). Therefore, P is equivalent to the linear programming problem

$$v(P_1) = \max_{(x,y) \in \text{conv} F(P)} (cx + dy). \qquad (P_1)$$

If $F(P_1) = \emptyset$, then $F(P) = \emptyset$ (case(a)). If $v(P_1) = \infty$, then $v(P) = \infty$ (case (b)). From Section 2.1 we know that if neither case (a) nor case (b) holds, then we have case (c), i.e., there exists an extreme point of $\text{conv} F(P)$ such that if $(x^*, y^*) \in F^*(P_1)$, then $(x^*, y^*) \in F^*(P)$.

If condition (3) of Theorem 3.1 is satisfied, $F(P)$ is the finite union of the polyhedral set of the form $\{(x', y) | Ax' + Dy = b, y \geqslant 0\}$, where x' in a nonnegative integer vector satisfying $\|x'\| < M$. If all such sets are empty, P is infeasible (case (a)). If $cx + dy$ is unbounded on one of such sets, P is unbounded (case(b)). Otherwise, $cx + dy$ is bounded on all such sets, and hence the set on which $cx + dy$ attains its maximum also contains the optimal solution of P. □

If $n = 0$, then P is a linear programming problem and Theorem 3.1 gives a sufficient condition for the trichotomy. To see why assumptions on P are needed, note that the equality-constrained problem obtained from Example 3.1 by the introduction of continuous slack variables will satisfy none of conditions (1)–(3), and does not satisfy the trichotomy. On the other hand, if in Example 3.1 we replace $x_1 \geqslant 1$ by $x_1 \geqslant 0$, the problem has the optimal solution $(0, 0)$, so none of the conditions of Theorem 3.1 are necessary conditions.

We note also that in Theorem 3.1 there is no constraint on vector b. It follows from that that $F(P)$ has a finite number of extreme points (Exercise 3.9.2). Consider a set

$$S = \{x | Hx \leqslant h, x \geqslant 0, x \in Z^n\}.$$

We show now that if H is a rational matrix, S has a finite number of extreme points. To do this, we find a positive integer k such that $A' = kH$ is an integer matrix. Then

$$S = \{x | A'x \leqslant \lfloor kh \rfloor, x \geqslant 0, x \in Z^n\}.$$

After adding the integer slack variables, $s \geqslant 0$, $s \in Z^m$, set S has the same form as the set in Exercise 3.9.2, namely

$$S = \{(x, s) | A'x + s = \lfloor kh \rfloor, (x, s) \geqslant 0, (x, s) \in Z^{n+m}\},$$

so S has a finite number of extreme points.

Let $\bar{P}$ denote the linear programming relaxation of P. Studying relations between $\bar{P}$ and P plays an important role in integer programming. As a start, we prove the following

THEOREM 3.2. *If* $F(\bar{P}) = \emptyset$, *then* $F(P) = \emptyset$. *If assumption 2 of Theorem 3.1 is satisfied and* $v(\bar{P}) = \infty$, *then either* $v(P) = \infty$ *or* $F(P) = \emptyset$. *If* $F^*(\bar{P}) \neq \emptyset$, *then either* $F^*(P) \neq \emptyset$ *or* $F^*(P) = \emptyset$. □

In practice the following corollary is useful.

COROLLARY 3.1. *If assumption 2 of Theorem 3.1 holds and* $F(P) \neq \emptyset$, *then* $v(P) = \infty$ *if and only if* $v(\bar{P}) = \infty$ *and* $F^*(P) \neq \emptyset$ *if and only if* $F^*(\bar{P}) \neq \emptyset$. □

In other words, if all data in P are rational and we know that P is feasible, after solving its linear programming relaxation $\bar{P}$ we know whether P has an optimal solution or not.

3.2. UNIMODULARITY

Consider a linear programming problem

$$v(\bar{P}) = \max \{cx | Ax = b, x \geqslant 0\}, \tag{$\bar{P}$}$$

and assume that all data in $\bar{P}$ are integers. We know (Section 2.1) that an optimal solution to $\bar{P}$ takes the form $\bar{x} = (B^{-1}b, 0)$, where B is an optimal basis of P. It is known that there exists such a matrix B^+ that

$$B^{-1} = \frac{B^+}{\det B}.$$

From the integrality of A follows the integrality of B^+. If $\det B = \pm 1$, then B^{-1} is an integer matrix and $\bar{x} = (B^{-1}b, 0)$ is an integer vector.

DEFINITION 3.1. A square, integer matrix B is called *unimodular* if $|\det B| = 1$. An integer matrix A is called *totally unimodular* if every square, nonsingular submatrix of A is unimodular. $\square$

Consider now the basic submatrices of A in problem P.

THEOREM 3.3. *If A is totally unimodular, every basic solution to P is integer.* $\square$

One may construct an example showing that Theorem 3.3 does not give a necessary condition for integer basic solutions for all integer vectors b. But if follows from Theorem 3.1 that if A is totally unimodular, it is sufficient to solve the linear programming relaxation of P, denoted as $\bar{P}$, and if $\bar{x} \in F^*(\bar{P})$, then $\bar{x} \in F^*(P)$.

We show now that for problems with inequality constraints

$$v(P') = \max\{cx | Ax \leqslant b, x \geqslant 0\}, \tag{P'}$$

total unimodularity of A is necessary and sufficient for all extreme points of $F(P')$ to be integers for every integer vector b.

THEOREM 3.4. *For any integer matrix A, the following statements are equivalent:*
1. *A is totally unimodular,*
2. *the extreme points (if any) of $S(b) = \{x | Ax \leqslant b, x \geqslant 0\}$ are integer for any integer b,*
3. *every square nonsingular submatrix of A has an integer inverse.*

Proof. $(1 \Rightarrow 2)$ After adding nonnegative slack variables, we have the system

$$Ax + Is = b, \quad x \geqslant 0, s \geqslant 0.$$

The extreme points of $S(b)$ correspond to basic solutions of the system. If a given basis B contains only columns from A, then x_B is integer as A is totally unimodular. The same is true if B contains only columns from I. So we have to consider the case when $B = (\bar{A}, \bar{I})$, where $\bar{A}$ is a submatrix of A and $\bar{I}$ is a submatrix of I. After the permutation of rows of B, we have

$$B' = \begin{bmatrix} A_1 & 0 \\ \hline A_2 & I' \end{bmatrix}.$$

Obviously, $\det B = \det B'$ and

$$|\det B'| = |\det A_1| |\det I'| = |\det A_1|.$$

Now A totally unimodular implies $|\det A_1| = 0$ or 1 and since B is assumed to be nonsingular, $|\det B'| = 1$ and x_B is an integer.

$(2 \Rightarrow 3)$. By the assumption, $x_B = B^{-1}b$ is an integer and it is sufficient to prove that $\bar{b}_j$ is an integer vector, where $\bar{b}_j$ is the jth column of B^{-1}, $j = 1, \ldots, m$. Let t be any integer vector such that $t + \bar{b}_j \geqslant 0$ and $b(t) = Bt + e_j$, where e_j is the jth unit vector. Then

$$x_B = B^{-1}b(t) = B^{-1}(Bt + e_j) = t + B^{-1}e_j = t + \bar{b}_j \geqslant 0,$$

so $(x_B, x_N) = (t + \bar{b}_j, 0)$ is an extreme point of $S(b(t))$. As x_B and t are integer vectors, $\bar{b}_j$ is an integer vector too for $j = 1, \ldots, m$ and B^{-1} is an integer.

$(3 \Rightarrow 1)$. Let B be an arbitrary square, nonsingular submatrix of A. Then

$$|\det BB^{-1}| = |\det B||\det B^{-1}| = 1.$$

By the assumption, B and B^{-1} are integer matrices. Thus $|\det B| = |\det B^{-1}| = 1$, and A is totally unimodular.   □

In general, it is not easy to tell by Definition 3.1 whether a given matrix consisting only of $(0, +1, -1)$ is totally unimodular. The theorem below gives a sufficient condition that is useful in many applications (see Sections 3.3 and 3.4).

THEOREM 3.5. *An integer matrix* $A = (a_{ij})$, $i = 1, \ldots, m$, $j = 1, \ldots, n$, *with all* $a_{ij} = 0, 1$ *or* -1 *is totally unimodular if*

1. *no more than two nonzero elements appear in each column,*
2. *the rows of A can be partitioned into two subsets M_1 and M_2 such that*

(a) *if a column contains two nonzero elements with the same sign, one element is in each of the subsets,*

(b) *if a column contains two nonzero elements of opposite signs, both elements are in the same subset.*

Proof. The proof is by induction. One element submatrix of A has a determinant equal to $(0, 1, -1)$. Assume that the theorem is true for all submatrices of A of order $k-1$ or less. Let B be any submatrix of A of the order k. If B contains a null vector, $\det B = 0$. If B contains a column with only one nonzero element, we expand $\det B$ by that column and apply the induction hypothesis. Finally, consider the case in which every column of B contains two nonzero elements. Then from 2(a) and 2(b) for every column j

$$\sum_{i \in M_1} b_{ij} = \sum_{i \in M_2} b_{ij}, \quad j = 1, \ldots, k.$$

Let b_i be the ith row. Then the above equality gives

$$\sum_{i \in M_1} b_i - \sum_{i \in M_2} b_i = 0,$$

which implies that $\det B = 0$.   □

In general, the conditions of Theorem 3.5 are not necessary for total unimodularity. However, it can be proved that for the class of matrices which satisfy condition 1, condition 2 is necessary. An elegant interpretation of Theorem 3.5 may be given using graph theory. Therefore, in the next section, we give some basic definitions of graph theory which we will use in this book.

3.3. BASIC DEFINITIONS OF GRAPH THEORY

A (finite) *graph* G is an ordered pair (V, E), where $V = \{1, \ldots, m\}$ is a set of *vertices* and $E \subseteq V \times V$ is a set of *edges* of G if we identify (i, j) with (j, i), otherwise (i, j) is called an *arc* (directed edge) of G.

We will say that an edge $e_k = (i, j)$ joins vertices i and j, while an arc $e_k = (i, j)$ is represented as a line from i to j with an arrowhead pointing to j. If $(i, j) \neq (j, i)$, the corresponding graph is called a directed one, otherwise it is an undirected one. An edge (i, i) is called a *loop*.

A graph then is defined if we specify V and E. Another way consists of using the so-called *incidence matrix*. Let $|V| = m$ and $|E| = n$. The incidence matrix $A = (a_{ij})$, $i = 1, \ldots, m$, $j = 1, \ldots, n$, of a given undirected graph is defined as:

$$a_{ij} = \begin{cases} 1 & \text{if edge } e_j \text{ is incident with vertex } i, \\ 0 & \text{otherwise.} \end{cases}$$

For a given directed graph, the incidence matrix is defined as follows:

$$a_{ij} = \begin{cases} -1, & \text{if} \quad e_j = (k, i), \quad k \neq i, \\ 1, & \text{if} \quad e_j = (i, k), \quad k \neq i, \\ 0, & \text{otherwise.} \end{cases}$$

Note that the incidence matrix of a given directed graph satisfies the assumptions of Theorem 3.5. Each column (arc) has at most two nonzero elements: one $(+1)$ and one (-1). Condition 2 of Theorem 3.5 is obviously fulfilled with $M_1 = \{1, \ldots, m\}$ and $M_2 = \emptyset$.

COROLLARY 3.2. *The incidence matrix of any directed graph is totally unimodular.* □

In an undirected graph, two edges are said to be *adjacent* if they are both incident to the same vertex. In a directed graph, $e_i = (h, j)$ and $e_k = (l, p)$ are adjacent if $j = l$. A sequence of distinct edges $t = (e_{j_1}, e_{j_2}, \ldots e_{j_p})$, where e_{j_k} and $e_{j_{k+1}}$ are adjacent, is called a *path*. A path is sometimes identified by its beginning and ending vertices; if $e_{j_1} = (i_0, i_1)$ and $e_{j_p} = (i_{p-1}, i_p)$, then t is said to be a path connecting i_0 and i_p. In a directed graph, the path is said to be from i_0 to i_p. A path that begins and ends at the same vertex is called a *cycle*. Paths without cycles are called *simple paths*.

A graph $G' = (V', E')$ is called a *partial graph* of $G = (V, E)$ if $V' \subseteq V$ and $E' \subseteq V' \times V'$. A partial graph with $V' = V$ is called a *subgraph* of G. An undirected graph is *connected* if for all $i, j \in V$, $i \neq j$, there exists at least one path connecting i and j. A directed graph is called *strongly connected* if for all $i, j \in V$, $i \neq j$, there exists a path from i to j and from j to i. A connected, undirected graph is said to be a *tree* if it contains no cycles. A subgraph of $G = (V, E)$ which is a tree is called a *spanning tree*. It can be shown that for a tree $G = (V, E)$, we have $|E| = |V| - 1$ (Exercise 3.9.5).

3.4. FLOWS IN NETWORKS

Using the definitions from Section 3.3 we will now formulate some integer programming problems connected with optimization of flows in networks. We show that the assumptions from Theorem 3.5 are fulfilled for these problems and they can be solved by the simplex method. However, due to the specific structure of these problems, computational methods have been developed which are more efficient than the simplex method.

3.4.1. *General Formulation*

Many commodities, such as gas, oil, etc., are transported through networks in which we distinguish sources, intermediate or distribution points and destination points. Many practical problems may be formulated in a similar way. More than thirty years of experience have proved the usefulness of such an approach. According to recent estimations, over 80% of all solved linear programming problems are network flow problems.

We will represent a network as a directed graph $G = (V, E)$ and associate with each arc $(i, j) \in E$ the flow x_{ij} of the commodity and the capacity d_{ij} (possibly infinite) that bounds the flow through the arc. The set V is partitioned into three sets: V_1 — set of sources or origins, V_2 — set of intermediate points, and V_3 — sets of destinations or sinks. For each $i \in V_1$, let a_i be a supply of the commodity and for each $i \in V_3$, let b_i be a demand for the commodity. We assume that there is no loss of the flow at intermediate points. Additionally, by $V(i)$ ($V'(i)$) we denote a set of arcs going out of (going into) the vertex i, i.e.,

$$V(i) = \{j | (i, j) \in E\} \quad \text{and} \quad V'(i) = \{j | (i, j) \in E\}.$$

The *minimum cost capacitated problem* may be formulated as:

$$v(P) = \min \sum_{\{i, j\} \in E} c_{ij} x_{ij}, \tag{P}$$

subject to

$$\sum_{j \in V(i)} x_{ij} - \sum_{j \in V'(i)} x_{ji} \begin{cases} \leqslant & a_i, & i \in V_1, \\ = & 0, & i \in V_2, \\ \leqslant & -b_i, & i \in V_3, \end{cases} \tag{3.1}$$

$$0 \leqslant x_{ij} \leqslant d_{ij}, \quad (i, j) \in E. \tag{3.2}$$

Constraint (3.1) requires the conservation of flow at intermediate points, a net flow into sinks at least as great as the demand, and a net flow out of sources equal or less than the supply. In some applications, demand must be satisfied exactly and all of the supply must be used. If all of the constraints of (3.1) are equalities, the problem has no feasible solution unless

$$\sum_{i \in V_1} a_i = \sum_{i \in V_3} b_i.$$

To avoid pathological cases, we assume for each cycle in the network $G = (V, E)$ either that the sum of costs of arcs in the cycle is positive or that the minimal capacity of an arc in the cycle is bounded. In the opposite case, P is unbounded, $v(P) = -\infty$ and $F^*(P) = \emptyset$.

THEOREM 3.6. *The constraint matrix corresponding to* (3.1) *and* (3.2) *is totally unimodular.*

Proof. The constraint matrix has the form

$$A = \begin{bmatrix} A_1 \\ I \end{bmatrix},$$

where A_1 is the matrix for (3.1) and I is an identity matrix for (3.2). In the proof of Theorem 3.4, we show that A totally unimodular implies (A, I) totally unimodular. Obviously, the transpose of a totally unimodular matrix is totally unimodular. Thus it suffices to establish the total unimodularity of A_1.

Each variable x_{ij} appears in exactly two constraints (3.1) with coefficients $+1$ and -1. Thus A_1 is an incidence matrix for a directed graph and therefore it satisfies the condition of Theorem 3.5 and by Corollary 3.2 is totally unimodular. ☐

3.4.2. *Some Particular Cases*

The most popular case of P is the so-called (capacitated) *transportation problem.* We obtain it if we put in P: $V_2 = \emptyset$, $V'(i) = \emptyset$ for all $i \in V_1$ and $V(i) = \emptyset$ for all $i \in V_3$. So we get

$$v(T) = \min \sum_{(i,j) \in E} c_{ij} x_{ij}, \tag{T}$$

$$\sum_{j \in V(i)} x_{ij} \leqslant a_i, \quad i \in V_1,$$

$$\sum_{j \in V'(i)} x_{ji} \geqslant b_i, \quad i \in V_3,$$

$$0 \leqslant x_{ij} \leqslant d_{ij}, \quad (i,j) \in E.$$

If $d_{ij} = \infty$ for all $(i, j) \in E$, the uncapacitated version of P is sometimes called the *transshipment problem.*

If all $a_i = 1$, all $b_i = 1$ and, additionally, $|V_1| = |V_2|$, the transshipment problem reduces to the so-called *assignment problem* of the form

$$v(AP) = \min \sum_{i \in V_1} \sum_{j \in V(i)} c_{ij} x_{ij}, \tag{AP}$$

$$\sum_{j \in V(i)} x_{ij} = 1, \quad i \in V_1,$$

$$\sum_{j \in V'(i)} x_{ji} = 1, \quad i \in V_3,$$

$$x_{ij} \geqslant 0.$$

Note that $|V_1| = |V_3|$ implies that all constraints in AP must be satisfied as equalities. In Section 1.5, we formulate AP as a binary problem. By Theorem 3.6, we know that we can solve AP as a linear programming problem, e.g., by the simplex method.

Let $V = \{1, \ldots, m\}$. Still another important practical problem obtained from P is called the *maximum flow problem*. In this problem, $V_1 = \{1\}$, $V_3 = \{m\}$, $V'(1) = \emptyset$, $V(m) = \emptyset$, $a_1 = \infty$, $b_m = \infty$. The problem is to maximize the total flow into the vertex m under the capacity constraints

$$v(MF) = \max \sum_{i \in V'(m)} x_{lm}, \qquad (MF)$$

$$\sum_{j \in V(i)} x_{ij} - \sum_{j \in V'(i)} x_{ji} = 0, \quad i \in V_2 = \{2, \ldots, m-1\},$$

$$0 \leqslant x_{ij} \leqslant d_{ij}, \quad (i,j) \in E.$$

Finally, consider the *shortest path problem*. Let c_{ij} be interpreted as the length of edge (i, j). Define the length of a path in G to be the sum of the edge lengths over all edges in the path. The objective is to find a path of minimum length from a vertex 1 to vertex m. It is assumed that all cycles have nonnegative length. This problem is a special case of the transshipment problem in which $V_1 = \{1\}$, $V_3 = \{m\}$, $a_1 = 1$ and $b_m = 1$. Clearly, the demand at vertex m will be satisfied with equality by exactly one unit of flow from vertex 1. The capacity constraints $d_{ij} = 1$ are implicitly accounted for, since $a_1 = 1$, and no negative cycles imply that there are optimal solutions in which $x_{ij}^* \leqslant 1$ for all $(i, j) \in E$. By Theorem 3.6, $x_{ij}^* = 0$ or 1 for all $(i, j) \in E$.

In Chapter 6, we show that the integer programming problem is equivalent to the shortest path problem in a specific graph with $c_{ij} \geqslant 0$ for all $(i, j) \in E$. Therefore, in the next subsection we consider Dijkstra's algorithm solving such a shortest path problem.

Let A be the incidence of the directed graph $G = (V, E)$, where $V = \{1, \ldots, m\}$ and $E = \{e_1, \ldots, e_n\}$. With each arc e_j we associate its length $c_j \geqslant 0$ and its flow $x_{ij} \geqslant 0$. The shortest path problem in the graph $G = (V, E)$ may be formulated as:

$$v(SP) = \min \sum_{j=1}^{n} c_j x_j, \qquad (SP)$$

$$Ax = \begin{bmatrix} +1 \\ 0 \\ \vdots \\ 0 \\ -1 \end{bmatrix}, \quad x \geqslant 0.$$

The first constraint corresponds to the source vertex, the mth constraint corresponds to the demand vertex, while the remaining constraints correspond to the intermediate vertices, i.e., the points of distribution of the unit flow.

The dual problem to SP is

$$v(DSP) = \max(u_1 - u_m), \qquad (DSP)$$

$$uA \leqslant c. \qquad (3.3)$$

Since in SP we have m equalities, in the dual we have m variables unrestricted in sign. The constraints (3.3) are equivalent to

$$u_i - u_j \leqslant c_{ij} \quad \text{for each} \quad (i, j) \in E. \tag{3.4}$$

From complementarity slackness (Section 2.6, Theorem 2.6) follows that $x^* \in R^n$ and $u^* \in R^m$ are optimal solutions to SP and DSP, respectively, if and only if

$$u_l^* - u_p^* = c_{lp} \quad \text{if} \quad e_j = (l, p) \in E \text{ and } x_j^* = 1$$

and

$$u_l^* - u_p^* < c_{lp} \quad \text{if} \quad e_j = (l, p) \in E \text{ and } x_j^* = 0.$$

The dual variable u_i is called the *label* of vertex i.

3.4.3. *Dijkstra's Method*

The method proposed by Dijkstra is an example of the so-called *primal dual methods* in which we solve the primal problem and the dual problem at the same time — in our case, SP and DSP. We know (Section 2.6) that the solution which is primal and dual feasible is the optimal one.

Without loss of generality, we may put $u_1 = 0$, $u_i = \infty$ for $i = 2, ..., m$, which gives the dual feasible solution, but not the primal one as by the complementarity slackness $x_j = 0$ for $j \in E$. The idea of Dijkstra's method consists in labelling of vertices (changing u_i for $i = 2, ..., m$), starting from vertex 1 preserving the dual feasibility in order to reach the primal feasibility. Doing this, we have smaller and smaller dual problems (the number of dual variables decreases), and when we reach vertex m, we have the dual and the primal feasibility, thus the solution constructed in such a way is optimal. Moreover, since DSP is very simple, there is no necessity of using the simplex tableous.

Let $W \subseteq V$ be a set of labelled vertices; at the beginning, $W = \{1\}$. By u_i we denote the length of the shortest path from 1 to i passing through vertices from W. At the beginning, we have $u_1 = 0$, $u_i = \infty$ for $i = 2, ..., m$. In each iteration of Dijkstra's algorithm, we select a minimum label element (vertex) $r \in V - W$ with the label u_r and add it to W. When $W = V$, we have found the shortest path having the length u_m.

The assumption $c_j \geqslant 0$, $j \in E$, is essential as then adding an arc will not decrease the length of the path constructed so far. Therefore, to remember the shortest path from 1 to i it is enough to remember the predecessor of i denoted as $p(i)$, $i = 2, ..., m$. Obviously, we may assume that if $e_j = (k, l) \notin E$, then $c_j = \infty$.

Dijkstra's Algorithm
Step 0 (Initialization): $u_1 := 0$; $p(1) := 0$; $W = \{1\}$; For all $i \in V - W$, set $u_i := c_{1i}$ and $p(i) := 0$;
Step 1 (Basic iteration): While $W \neq V$, do (a) and (b);
  (a) (Updating W): Find $u_r = \min\{u_i | i \in V - W\}$; $\quad W := W \cup \{r\}$;
  (b) (Labelling): For $i \in V - W$, if $u_i > u_r + c_{ri}$, then $(u_i := u_r + c_{ri}$, and $p(i) := r)$;
  Stop. $\qquad\qquad\qquad\qquad\qquad\qquad\qquad\qquad\qquad\qquad\qquad\qquad\quad \square$

Now we estimate the number of additions and comparisons in Dijkstra's algorithm. At the kth basic iteration, $k = 1, \ldots, m-1$, we are doing $m-k$ additions and $m-k$ comparisons in labelling u_i for $i \in V - W$. Additionally, we have to compare $m-k-1$ numbers to find vertex r. So finally, to find the shortest path from 1 to m in a graph $G = (V, E)$ we have to do $O(m^2)$ additions and comparisons, where $m = |V|$.

COROLLARY 3.3. *Dijkstra's algorithm has the complexity* $O(|V|^2)$. $\square$

Example 3.2. Find the shortest path from 1 to 6 on the graph in Fig. 3.2. Figures above arrows indicate the length of arcs.

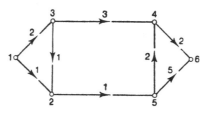

Fig. 3.2.

At Step 0, we have $u_1 = 0$, $u_2 = 1$, $u_3 = 2$ and $u_i = \infty$ otherwise. At Step 1(a), we find $u_2 = \min\{u_i | i \in V - \{1\}\}$ and $W = \{1, 2\}$. Step 2(b) gives $u_5 = 2$, $p(5) = 2$ and the remaining u_i and $p(i)$ do not change.

At the next iteration, we choose $r = 3$, and then $u_4 = 5$ and $W = \{1, 2, 3\}$. In the same way, in the next iterations, we get $W = \{1, 2, 3, 5\}$, $W = \{1, 2, 3, 5, 4\}$ and $W = \{1, 2, 3, 5, 4, 6\}$, which gives $u_6 = 6$ and $p(6) = 4$. So we get the shortest path passing through 2, 5, 4 (Exercise 3.9.8). $\square$

By Dijkstra's algorithm, we may find the shortest path in the sense of graph theory. To do this, we put $c_j = 1$, $j \in E$, and $c_j = \infty$ if $j \notin E$ (Exercise 3.9.9).

Dijkstra's algorithm also has one interesting feature, namely it find the shortest path from vertex 1 to any other vertex $i = 2, \ldots, m$. The labels u_i obtained at the last iterations are the lengths of such paths. If we want to compute the shortest path between any pair of vertices, we have to repeat Dijkstra's algorithm taking as the vertex 1 the vertex i, $i = 2, \ldots, m$. Thus computing the shortest path for any pair of vertices in $G = (V, E)$ requires $O(|V|^3)$ operations.

3.5. A FUNDAMENTAL CUT

Consider the integer programming problem of the form

$$v(P) = \max\{cx | Ax = b, x \geqslant 0, x \in Z^n\}. \tag{P}$$

If an optimal solution to its linear programming relaxation, i.e., to the problem

$$v(\bar{P}) = v(\bar{P}_0) = \max\{cx | Ax = b, x \geqslant 0\}, \tag{$\bar{P}$}$$

is not an integer, then P can be solved by one of the so-called *cutting-plane methods*.

The idea of any cutting plane method consists in adding the constraint (cut) $\alpha x \leqslant \beta$ which cuts off (separates) the optimal solution to $\bar{P}_0$ from $F(P)$ preserving $F(P)$. As a result, we obtain the linear programming problem $\bar{P}_1$. If its optimal solution is an integer, it is at the same time an optimal solution to P. Otherwise, we add a new cut to $\bar{P}_1$ obtaining $\bar{P}_2$ and so on. In the next section, we show that under mild assumptions the sequence $P_0, P_1, \ldots$ is finite, i.e., there exists an index r such that $\bar{x} \in F^*(\bar{P}_r)$ is an integer. In this section, we derive such a cut.

We know (Section 2.2) that an optimal solution to $\bar{P}$ may be written in the form of the simplex tableau

$$x_{Bi} = h_{i0} - \sum_{j \in R} h_{ij} x_j, \quad i = 0, 1, \ldots, m, \tag{3.5}$$

where x_{Bi} is the ith basic variable and h_{i0} is its value, $i = 1, \ldots, m$. The zero row corresponds to the objective function. By R we denote the index set of nonbasic variables.

Multiplying the ith row of (3.5) by some number $p \neq 0$, called the *cut parameter*, we get

$$px_{Bi} + \sum_{j \in R} ph_{ij} x_j = ph_{i0}.$$

As in P we have $x \geqslant 0$,

$$\lfloor px_{Bi} \rfloor + \sum_{j \in R} \lfloor ph_{ij} x_j \rfloor \leqslant ph_{i0}.$$

Since the left-hand side of the above inequality is an integer, it cannot exceed the integer part of the right-hand side, which implies that

$$\lfloor p \rfloor x_{Bi} + \sum_{j \in R} \lfloor ph_{ij} \rfloor x_j \leqslant \lfloor ph_{i0} \rfloor. \tag{3.6}$$

Multiplying the same ith constraint of (3.5) by $\lfloor p \rfloor$ and subtracting from it (3.6) we obtain

$$\sum_{j \in R} (\lfloor p \rfloor h_{ij} - \lfloor ph_{ij} \rfloor) x_j \geqslant \lfloor p \rfloor h_{i0} - \lfloor ph_{i0} \rfloor. \tag{3.7}$$

Expression (3.7) is a general formula for the cut generated from the ith row of the simplex tableau. The ith row is then called the *source row*. For different p, we get from (3.7) different cuts and develop different cutting plane algorithms. For instance, if p is an integer, we obtain the so-called *method of integer forms*. In particular, for $p = 1$, by (3.7), we get

$$\sum_{j \in R} (h_{ij} - \lfloor h_{ij} \rfloor) x_j \geqslant h_{i0} - \lfloor h_{i0} \rfloor.$$

Letting $h_{ij} = \lfloor h_{ij} \rfloor + f_{ij}$ we obtain

$$\sum_{j \in R} f_{ij} x_j \geqslant f_{i0}, \tag{3.8}$$

This is the cut in the method of integer forms for $p = 1$.

Let (3.5) be an optimal simplex tableau for $\bar{P}$, so we have a solution which is primal and dual feasible. If all $x_{Bi} = h_{i0}$ are integers, then (x_B, x_N) with $x_N = 0$ is, at the same time, an optimal solution for the integer programming problem P. If it is not the case, there exists at least one row of (3.5), say the rth row of (3.5), for which $x_{Br} = \lfloor h_{r0} \rfloor + f_{r0}$ with $f_{r0} > 0$ and $r > 1$. The case when the above is true only for the zero row of (3.5) is impossible as

$$h_{00} = \sum_{i=1}^{m} c_{Bi} h_{i0}.$$

If $h_{i0} \in Z$ for $i = 1, \ldots, m$, by the integrality of data, $h_{00} \in Z$ too.

If $f_{r0} > 0$, the optimal basic solution to $\bar{P}$ violates (3.8). In other words, (3.8) cuts off $\bar{x} \in F^*(\bar{P})$. Furthermore, (3.7) and (3.8) do not exclude any feasible integer solutions, since, as shown above, they are implied by integrality.

Adding a slack variable $s \geqslant 0$ we may write (3.8) as an equality

$$s = -f_{i0} - \sum_{j \in R} f_{ij}(-x_j). \tag{3.9}$$

Note that $s \in Z$, as by (3.5)

$$x_{Bi} = -\left(-f_{i0} + \sum_{j \in R} f_{ij} x_j\right) + \left(\lfloor h_{i0} \rfloor - \sum_{j \in R} \lfloor h_{ij} \rfloor x_j\right),$$

and the expression in the second bracket is an integer.

Adding (3.9) to P we obtain problem P_1 with $m+1$ constraints and $n+1$ integer variables. If we add (3.9) to the simplex tableau (3.5), we have a dual feasible solution which is not primal feasible, as $s < 0$. The primal feasibility may be reached by the dual simplex algorithm (Section 2.7). It may take more than one simplex iteration.

Algorithm of Integer Forms
Step 0 (Initialization): Solve $\bar{P}$; Let $\bar{x} \in F^*(\bar{P})$;
Step 1 (Optimality test): Is $\bar{x}$ an integer? If so, $\bar{x} \in F^*(P)$ and stop;
Step 2 (Cut generation): Choose the source row; Add the cut (3.9) to the simplex tableau (3.5);
Step 3 (Reoptimalization): Apply the dual simplex algorithm to the simplex tableau (3.5); If $h_{00} = -\infty$, stop $(F(P) = \emptyset)$, otherwise let $\bar{x}$ be an optima solution; Go to Step 1.   □

Example 3.3. We solve the problem from Example 1.1 by the method of integer forms. Table (2.5) is optimal for this problem. We give it here once more

TABLE 3.1

		$-x_3$	$-x_4$
x_0	38/7	4/7	1/7
x_1	11/7	3/7	$-1/7$
x_2	16/7	2/7	3/7

By (3.8) we get

$$\tfrac{4}{7}x_3 + \tfrac{1}{7}x_4 \geqslant \tfrac{3}{7} \quad (\text{row } x_0),$$

$$\tfrac{3}{7}x_3 + \tfrac{6}{7}x_4 \geqslant \tfrac{4}{7} \quad (\text{row } x_1),$$

$$\tfrac{2}{7}x_3 + \tfrac{3}{7}x_4 \geqslant \tfrac{2}{7} \quad (\text{row } x_2).$$

We arbitrarily choose as a cut the inequality corresponding to the zero row. Writing it in the form (3.9) we get

$$s_1 = -\tfrac{3}{7} - \tfrac{4}{7}(-x_3) - \tfrac{1}{7}(-x_4),$$

which gives the simplex tableau of Table 3.2

TABLE 3.2

		$-x_3$	$-x_4$
x_0	38/7	4/7	1/7
x_1	11/7	3/7	$-1/7$
x_3	16/7	2/7	3/7
s_1	$-3/7$	$-4/7$	$-1/7$

This tableau is dual feasible, but it is not primal feasible. By the dual simplex algorithm, s_1 is leaving the basis and x_4 is entering the basis. After one iteration, we get the simplex tableau (see Table 3.3) which is primal and dual feasible. Moreover, this basic optimal solution is an integer, thus $x^* = (2, 1)$ is an optimal solution to P.

TABLE 3.3

		$-x_3$	$-s_1$
x_0	5	0	1
x_1	2	1	-1
x_2	1	$-10/7$	3
x_4	3	4	-7

One can easily check that if we choose the inequality corresponding to x_2 as a cut, we have to do at least one iteration more, as is shown in Fig. 1.3.

We write now the cut corresponding to row x_0 using the variables x_1 and x_2. We get

$$x_3 = 7 - 3x_1 - x_2,$$

$$x_4 = 10 - 2x_1 - 3x_2.$$

Substituting them into the cut we obtain

$$2x_1 + x_2 \leqslant 5$$

(compare Fig. 1.3).   □

3.6. FINITENESS OF THE METHOD OF INTEGER FORMS

In this section, we prove that the algorithm of integer form is finite, i.e., after adding of a finite number of cuts, an optimal solution to the corresponding linear programming problem is an integer. In the proof, we will use the lexicographic dual simplex algorithm.

We know (see Section 2.7) that the variable x_k is entering the basis if

$$\frac{h_{0k}}{h_{rk}} = \max_{j \in R} \left\{ \frac{h_{0j}}{h_{rj}} | h_{rj} < 0 \right\}, \tag{3.10}$$

where x_{Br} is leaving the basis, i.e., $h_{r0} < 0$ and usually

$$h_{r0} = \min_{i=1,\dots,m} h_{i0}.$$

Obviously, the choice of x_k in (3.10) is, in general, not unique, and therefore, similarly as in the primal simplex algorithm, we introduce the lexicographic criterion of the choice of x_k

$$\frac{1}{h_{rk}} h_k = \operatorname{lex\,max}_{j \in R} \left\{ \frac{1}{h_{rj}} h_j | h_{rj} < 0 \right\}, \tag{3.11}$$

where h_j is the jth column of the simplex tableau (3.5). The choice in (3.11) is unique. By putting cut rows at the bottom of the tableau, the lexicographic order of columns is not disturbed.

THEOREM 3.7. *The algorithm of integer forms in which*:

(a) *the first row x_{Br} in (3.5) with $f_{r0} > 0$ is chosen as a source row at Step 1 of the algorithm and*

(b) *the lexicographic dual simplex algorithm is used at the Step 3 of the algorithm*

is finite, i.e., after adding of a finite number of cuts, either $F(P) = \varnothing$ or we find an optimal solution to P.

Proof. Without loss of generality, we may assume that $\bar{P}$ is bounded, e.g., we may add the extra constraint

$$\sum_{j=1}^{n} x_j \leqslant M, \tag{3.12}$$

where M is a sufficiently large number. The constraint (3.12) does not change the lexicographic order of columns in (3.5) as in (3.12) all coefficients are positive. Thus each linear programming problem solved by the method of integer forms is bounded.

Consider the rth iteration of the algorithm of integer forms. If at Step 3 we find that the dual problem is unbounded, problem $\bar{P}_r$ is inconsistent and in consequence P is inconsistent.

We will show now that the zero columns of (3.5) obtained in the subsequent iterations of the algorithm form the sequence

$$h_0^1 >_L h_0^2 >_L h_0^3 >_L \dots$$

To see this, consider problem $\bar{P}_r$. Since it is bounded, $h_{00}^r = \lfloor h_{00}^r \rfloor + f_{00}^r$. Let $f_{00}^r > 0$, then by (a) the zero row is a source row. By (3.9) we get the cut

$$s = -f_{00}^r + \sum_{j \in R} f_{0j}^r x_j .$$

The variable s becomes basic and let by (3.11) x_k enter the basis. The value of the objective function then will be

$$h_{00}^{r+1} = h_{00}^r - \frac{h_{0k}^r}{f_{0k}^r} \cdot f_{00}^r . \tag{3.13}$$

As the tableau (3.5) is optimal for $\bar{P}_r$, we have $0 < f_{0k}^r \leqslant h_{0k}^r$. Therefore, after one iteration of the dual simplex algorithm,

$$h_{00}^{r+1} < h_{00}^r - f_{00}^r = \lfloor h_{00}^r \rfloor .$$

The same consideration may be repeated for the first, the second and so on row of (3.5). Since $x_B \geqslant 0$, after a finite number of iterations of the lexicographic dual simplex algorithm, we obtain that the zero column of (3.5) is an integer, i.e., it is an optimal solution of P. $\qquad \square$

By Theorem 3.7, one may estimate the maximal number of iterations needed to solve any integer problem P with m constraints and n variables (see Section 3.8 for references).

3.7. OTHER MODIFICATIONS OF THE CUTTING-PLANE METHOD

The algorithm of integer forms has two drawbacks.

First, the coefficients in (3.9) are fractional which is represented by their approximation, and which may lead to the cumulation of numerical errors. The other source of errors is the integrality test. In computer calculation, we assume that a real number h is an integer (equal one) if $\min\{1-f, f\} < \varepsilon$, where ε is a given positive constant and f is a fractional part of h. Failing to recognize an integer may lead to unnecessary iterations, invalid cuts and even to loss of an optimal solution. Conversely, the improper identification of an integer could cause false termination. If it is desired to avoid the roundoff and integrality test problems completely, all calculations must be done with integers. The dual all-integer algorithm has this property.

Second, the method of integer forms uses the dual simplex algorithm, which means that the first feasible solution is an optimal one. Therefore, if we stop calculations before the first feasible solution is found, we have no optimal solution nor any feasible solution as an approximation of it. This drawback is removed by the primal all-integer algorithm.

3.7.1. Dual All-Integer Cuts

The coefficients in the dual all-integer cut are integers. It can be obtained if we choose p in (3.7) in the appropriate way. Since data in P are integers, we may modify the simplex method in such a way that all coefficients in any simplex tableau are integers.

Consider a dual feasible basic solution determined from a given all-integer tableau (3.5). If this solution is also primal feasible, it is optimal both to $\bar{P}$ and to P, so we assume that it is not. We can drive toward primal feasibility using the dual simplex algorithm. If the magnitude of the pivot element for each dual simplex iteration equals 1, the tableau will remain all-integer. Then, if primal feasibility is achieved, the solution will be optimal. If, by chance, there is $h_{i0} < 0$ and a dual pivot element of -1, an ordinary dual simplex iteration can be executed. Generally, this will not be the case.

By our assumption, we have in (3.5) at least one $x_{Br} = h_{r0} < 0$. In particular, it can correspond to the cut (3.7). If we choose p to be such that the pivot element equals -1, after the simplex iteration, we will again have an all-integer tableau. For $0 < p < 1$, after introducing a slack variable $s \geq 0$, by (3.7) we have

$$s = \lfloor ph_{r0} \rfloor - \sum_{j \in R} \lfloor ph_{rj} \rfloor x_j. \tag{3.14}$$

Since $h_{r0} < 0$ and $p > 0$, then $\lfloor ph_{r0} \rfloor \leq -1$ in (3.14).

Let $R_r = \{j | h_{rj} < 0, j \in R\}$ be a index set of variables which may enter the basis in the dual simplex algorithm. Our aim is the selection of x_k, $k \in R_r$, and p in such a way that $\lfloor ph_{rk} \rfloor = -1$ and dual feasibility is preserved, i.e.,

$$h_{0j} - \frac{\lfloor ph_{rj} h_{0k} \rfloor}{\lfloor ph_{rk} \rfloor} = h_{0j} + \lfloor ph_{rj} \rfloor h_{0k} \geq 0 \tag{3.15}$$

for all $k \in R_r - \{r\}$. Then

$$\lfloor ph_{rj} \rfloor \geq -\lfloor h_{0j}/h_{0k} \rfloor.$$

As $h_{rj} < 0$, we have

$$ph_{rj} \geq -\lfloor h_{0j}/h_{0k} \rfloor \quad \text{for all } j \in R_r - \{k\}.$$

Therefore, we get

$$p \leq \min_{j \in R_r - \{k\}} \frac{\lfloor h_{0j}/h_{0k} \rfloor}{|h_{rj}|} \equiv p^*. \tag{3.16}$$

The choice of $p = p^*$ gives the largest decrease in the objective function, although we may choose p such that $p \leq p^*$ and $0 < p < 1$.

Since in (3.15) we have $h_{rj} < 0$, thus $\lfloor ph_{rj} \rfloor \leq -1$ and therefore the variable entering the basis has to satisfy

$$h_{0k} = \min_{j \in R_r} h_{0j}. \tag{3.17}$$

In the *dual all-integer algorithm*, we write the cut (3.14) as the last constraint in the all-integer tableau (3.5), where p is defined by (3.16).

3.7.2. *Primal All-Integer Cuts*

In many practical problems, we know a feasible solution and may write the simplex all-integer tableau which is primal feasible, but, in general, it is not dual feasible. If the pivot element in the primal simplex algorithm is 1, a primal feasible, all-integer solution can be maintained. The *primal all-integer algorithm* bases on the above idea.

Let (3.5) be an all-integer simplex tableau in which $h_{i0} \geqslant 0$, $i = 1, \ldots, m$, and for some k, we have $h_{0k} < 0$. If

$$\frac{h_{r0}}{h_{rk}} = \min_{i=1,\ldots,m} \left\{ \frac{h_{i0}}{h_{ik}} \,\middle|\, h_{ik} \geqslant 1 \right\}, \tag{3.18}$$

then by the primal simplex algorithm x_k enters the basis and x_{Br} leaves the basis. Note that $h_{ik} \geqslant 1$ because the tableau is all-integer. If $h_{rk} = 1$, an ordinary simplex iteration can be executed. If $h_{rk} > 1$, let $p = 1/h_{rk} < 1$ in (3.7), which, after adding a slack variable $s \geqslant 0$ and integer, gives

$$s = \lfloor h_{r0}/h_{rk} \rfloor - \sum_{j \in R} \lfloor h_{rj}/h_{rk} \rfloor x_j. \tag{3.19}$$

The coefficient of x_k in (3.19) equals 1. The cut (3.19) does not exclude the current feasible integer solution (it may be optimal) since $s \geqslant 0$ in (3.19). But, unless $h_{r0}/h_{rk} = \lfloor h_{r0}/h_{rk} \rfloor$, the solution that would have been obtained by entering x_k and removing x_{Br} is eliminated. To see this, suppose we pivot on h_{rk}. In the new solution, we have $x_j = 0$ for all $j \in R - \{k\}$ and $x_k = h_{r0}/h_{rk}$. Substituting this solution into (3.19) yields

$$s = \lfloor h_{r0}/h_{rk} \rfloor - h_{r0}/h_{rk} \leqslant 0,$$

and now $s = 0$ if and only if h_{r0}/h_{rk} is an integer. If this is the case, the new basic feasible solution is

$$x_{Bi} = h_{0i} - (h_{r0}/h_{rk}) h_{ik}, \quad i = 1, \ldots, m, i \neq r,$$

and $x_{Br} = h_{r0}/h_{rk}$. This solution is integer and feasible. If this is not the case, i.e. when $s < 0$, (3.19) cuts off this new basic solution.

3.7.3. *Stronger Cuts*

From the above consideration we have that a tableau (3.5) gives an optimal solution to P if the following three conditions hold:
  (i) Primal feasibility, $h_{i0} \geqslant 0$, $i = 1, \ldots, m$.
  (ii) Dual feasibility, $h_{0j} \geqslant 0$, for all $j \in R$.
  (iii) Integrality, $h_{i0} \in Z$, $i = 1, \ldots, m$.
  Each of the basic three algorithms maintains two of the three conditions and adds cuts (3.7) in an attempt to satisfy the third. In the algorithm of integer forms, the first two conditions hold, and by addition of cuts (3.8) we obtain integrality of the basic solution (condition (iii)). In the dual all-integer algorithm, conditions (ii) and (iii) hold, and by cuts (3.14) we satisfy condition (i), while in the primal

all-integer algorithm cuts (3.19) allow condition (ii) to be satisfied preserving conditions (i) and (iii).

Obviously, conditions (i) and (iii) are necessary, while (i)–(iii) are sufficient but not necessary. Now we will describe a general cutting-plane algorithm.

Cutting-Plane Algorithm

Step 0 (Initialization): Find a solution which satisfies two of the three conditions;

Step 1 (Optimality test): If the third condition is satisfied, terminate;

Step 2 (Reoptimalization): Add a cut (3.7) with p chosen appropriately as the last
   constraint in (3.5); Pivot to maintain the specified two conditions (more than
   one pivot may be necessary); Go to Step 1.   □

For different values p in (3.7), we obtain different cuts and algorithms of different efficiency. The efficiency of algorithms depends on, generally speaking, how large part of the set $F(\bar{P})-F(P)$ a given cut cuts off from $F(\bar{P})$.

We say that a given cut

$$\sum_{j \in R} d_j x_j \geqslant d_0$$

is *stronger* (or *implies*) than an inequality

$$\sum_{j \in R} q_j x_j \geqslant q_0$$

if $d_j \leqslant q_j$ for all $j \in R$, $d_0 \geqslant q_0$, and at least one inequality is strict.

By this definition, a facet of conv $F(P)$ is the strongest cut where a facet of conv $F(P)$ is defined as inequality in R^n of the form

$$\sum_{j=1}^{n} d_j x_j \geqslant d_0,$$

which is satisfied by any $x \in F(P)$, and, moreover, n linear independent $x \in F(P)$ satisfy it as an equality. We will study the properties of facets in Section 5.5. In Chapter 6, we will consider methods for construction of stronger cuts, while in the next section we will study the use of cuts in branch-and-bound methods.

Computational experiments show that subsequent cuts are increasingly more parallel. This may lead to ill-conditioned bases (Section 2.7) and to a cumulation of numerical errors. Roundoff errors are particularly serious in to method of integer form.

3.8. BIBLIOGRAPHIC NOTES

3.1. The conditions for the trichotomy of Theorem 3.1 were studied by Meyer in 1974. Gould and Rubin (1973) show that an integer programming problem with real data is equivalent to an integer programming problem with integer data.

3.2. Hoffman and Kruskal (1958) proved Theorem 3.2 and some modifications of it were considered by Veinott and Dantzig (1968). See also Chandrasekaran (1969) and Truemper (1978).

3.3. There are many books on graph theory: Berge (1962), Harary (1969), Gondran and Minoux (1979), Deo (1974).

3.4. The book by Ford and Fulkerson (1962) is basic in theory of flows in networks. Computational complexity of network problems is studied in the book by Lawler (1976) and in the book by Papadimitriou and Steiglitz (1982). The shortest path algorithm (Section 3.4) was developed by Dijkstra (1959).

3.5–3.7. The method of integer forms is developed in Gomory (1963a, b), although the idea of cutting plane methods is described in his papers from 1958. Dantzig, Fulkerson and Johnson (1954) used cut to solve the travelling salesman problem. Gomory (1963a) gives a description of the method of integer forms and the proof of its finiteness. Glover (1965c) and Gomory (1963b) give descriptions of the dual all-integer algorithm (see also Glover, 1965a, b, 1967).

The primal all-integer algorithm is described in Young (1968) and Glover (1968a). The construction of stronger cuts is studied in many papers, for instance, in Bowman and Nemhauser (1971), Rubin and Graves (1972), Jeroslow (1977, 1978). Kolokolov (1976) estimated the maximal number of cuts needed to solve P by the method of integer forms.

3.9. EXERCISES

3.9.1. Prove the existence of a sequence $(x_1^{(k)}, x_2^{(k)})$ in Example 3.1.

3.9.2. Show that in case 1 of Theorem 3.1, $F(P)$ has a finite number of extreme points (Meyer, 1974).

3.9.3. Prove that in case 2 of Theorem 3.1, conv $F(P)$ is a convex polyhedra (Meyer, 1974).

3.9.4. Prove Theorem 3.2.

3.9.5. Let $G = (V, E)$ be an undirected, connected graph. Prove that the following statements are equivalent: (a) G is a tree; (b) $|E| = |V|-1$; (c) there exists exactly one path for each pair of vertices of G; (d) after removing of an edge, G becomes unconnected.

3.9.6. Prove that for the class of matrices which satisfy condition 1 of Theorem 3.5, condition 2 is necessary.

3.9.7. Show that the formulations of the assignment problem given in Sections 1.5 and 3.4 are equivalent.

3.9.8. For the problem from Example 3.2, write the primal and the dual problems and solve them. Compare the steps of the simplex algorithm with Dijkstra's algorithm.

3.9.9. Modify Dijkstra's algorithm for the case $c_{ij} = 1$, $(i, j) \in E$.

3.9.10. Solve Example 3.3 choosing as a source row, first the row corresponding to x_1 and next to x_2. Compare the number of cuts.

3.9.11. Give a description of the dual all-integer algorithm.

3.9.12. Solve Example 3.3 by the dual all-integer algorithm.

3.9.13. Give a description of the primal all-integer algorithm and solve Example 3.3 by it.

3.9.14. Derive a formula for a cut in a mixed integer programming problem.

3.9.15. Show that a facet of $\operatorname{conv} F(P)$ is the strongest cut.

CHAPTER 4

Branch-and-Bound Methods

These methods are commonly considered as the most efficient tools for solving integer programming problems. In Section 4.1, we describe the idea of the branch-and-bound methods and introduce the basic relations. Any branch-and-bound method, as we mentioned in Section 1.4, consists of two basic procedures: branching or partitioning of the feasible solution set into some number of subsets and bounding or estimating of the optimal value of the objective function on these subsets. In this chapter, we show how linear programming may be used in such an estimation, while the estimations based on Lagrangean relaxations of a given problem are studied in Chapter 7.

Relatively good estimations are obtained via a surrogate problem, which in fact is a knapsack problem. Therefore, in Chapter 5, we consider methods for solving the knapsack problem.

Results of computational experiments show that reformulation of a given integer programming problem is a useful approach. Therefore, in Chapter 6, we consider different methods of obtaining the co-called tighter equivalent formulations for a given integer programming problem.

Thus Chapters 4–7 together form a full description of the branch-and-bound methods. In Chapter 10, we consider some practical aspects of solving a given integer programming problem.

4.1. THE IDEA OF A BRANCH-AND-BOUND METHOD

Consider an integer programming problem

$$v(P) = \min \{cx | Ax \geqslant b, x \geqslant 0, x_j \in Z, j \in N_c\}, \qquad (P)$$

where $N_c \subseteq N$ is an index set of integer variables and N is an index set of all variables. If $N = N_c$, then P is a (pure) integer programming problem and in general ($N \neq N_c$) P is a mixed integer programming problem. For $N_c = \emptyset$, we have a linear programming problem.

Without loss of generality, we may assume that P is bounded. In many practical problems, bounds of the type $0 \leqslant x_j \leqslant d_j, j \in N$, are given in advance or may be easily calculated. If this is not the case, we may add one more constraint

$$\sum_{j \in N} x_j \leqslant M,$$

where M is a sufficiently large positive number. If for a given optimal solution x^* of P we have

$$\sum_{j\in N} x_j^* = M,$$

then P is unbounded, $v(P) = -\infty$ and $F^*(P) = \emptyset$.

4.1.1. Terminology

In any branch-and-bound method, we first partition $F(P)$ into a finite number, say r, of subsets, which is equivalent to partitioning of P into r subproblems $PP_1, PP_2, \ldots, PP_r$, which form the so-called *list of subproblems*. So

$$v(P) = \min_{i=1,\ldots,r} v(PP_i), \quad \text{where} \quad v(PP_i) = \min_{x\in F(PP_i)} cx \tag{4.1}$$

and

$$F(P) = \bigcup_{i=1}^{r} F(PP_i), \quad F(PP_i)\cap F(PP_j) = \emptyset \quad \text{for } i \neq j. \tag{4.2}$$

Thus, instead of solving one problem P, we have to solve r subproblems which, upon the appropriate choice of the partition (4.2), may be easier to solve. Moreover, in a branch-and-bound method we do not solve PP_i but only find a lower bound (in the case of maximization, upper bound) on $v(PP_i)$. This may be done, e.g., by solving the linear programming relaxation of PP_i denoted as $\overline{PP_i}$, $i = 1, \ldots, r$. Since

$$v(\overline{PP_i}) \leqslant v(PP_i) \quad \text{for} \quad i = 1, \ldots, r, \tag{4.3}$$

$v(\overline{PP_i})$ is a lower bound on $v(PP_i)$. This is true even if $F(PP_i) = \emptyset$ since then we have $v(PP_i) = +\infty$.

Let x^* be an *incumbent*, i.e., the best feasible solution found so far. In many problems, x^* may be found by near-optimal methods (Chapter 9) or obtained from practice. Its value

$$z^* = \sum_{j\in N} c_j x_j^*$$

is called a *cutting value*. If we do not know $x^* \in F(P)$, e.g., at the beginning of computation, we assume that $z^* = \infty$.

Having $v(\overline{PP_i})$ and z^*, which, for the moment, we assume is finite, we may answer the question whether the consideration of PP_i is perspective or not.
If

$$v(\overline{PP_i}) \geqslant z^*, \tag{4.4}$$

then no feasible solution of PP_i has an objective function value better than $z^* = cx^*$. The case $F(PP_i) = \emptyset$ is also covered by (4.4) as then we have $v(PP_i) = \infty$. If (4.4) holds, the analysis of PP_i is not perspective since it cannot give a feasible solution to P with the value better than z^*. We will say that in such a case PP_i is *fathomed* and (4.4) is the first criteria for fathoming of subproblems.

We partition the case $v(PP_i) < z^*$ in two subcases.

(a) If

$$\bar{x} \in F^*(\overline{PP_i}) \quad \text{and} \quad \bar{x} \in F(P), \tag{4.5}$$

then the analysis of PP_i cannot give a solution better than $\bar{x}$, and PP_i is fathomed. We call (4.5) the second criteria for fathoming. As $c\bar{x} < z^*$, we have a new incumbent $x^* = \bar{x}$ and a new cutting value $z^* = c\bar{x}$.

(b) If

$$\bar{x} \in F^*(\overline{PP_i}), \quad \text{but} \quad \bar{x} \notin F(P), \tag{4.6}$$

then the consideration of PP_i is perspective and it may produce a feasible solution better than the incumbent x^*. Then we may proceed with PP_i exactly in the same way as with P, i.e., construct its subproblems, add them to the list of subproblems and compute bounds for them.

Since (4.2) is a partitioning of $F(P)$, the subsequent branchings are simply partitions of some parts of $F(P)$. Therefore, the above branch-and-bound method is finite since after a finite number of partitions, even if no subproblem would be fathomed by (4.4) and (4.5), we obtain subproblems which are linear programming problems. All such subproblems are fathomed by (4.5) as for them $v(PP_i) = v(\overline{PP_i})$ for all i. Obviously, the list of subproblems will then be empty. Thus we proved

THEOREM 4.1. *If P is bounded, then after a finite number of branchings, the subproblem list is empty. Then either $z^* = \infty$ and $F^*(P) = \emptyset$ or $z^* < \infty$ and $x^* \in F^*(P)$, i.e., the incumbent is an optimal solution.* $\qquad\square$

It is useful to represent a computation process by means of the so-called *tree of subproblems*. The root of the tree corresponds to the problem P, while the other vertices of the tree represent subproblems. We have an arc from P_i to P_j if and only if the problem P_j results from a partition of the problem P_i. There are no arcs going out from fathomed vertices. They are underlined in Fig. 4.1.

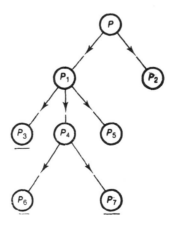

Fig. 4.1.

A typical tree of subproblems is presented in Fig. 4.1. We partition a given problem P into two subproblems P_1 and P_2, and at the beginning we have $z^* = \infty$. Since $v(\bar{P}) < z^*$, the consideration of P_1 is perspective.

Let us assume that we branch P_1 into 3 subproblems P_3, P_4 and P_5, and assume that we have found $x^* \in F(P_1)$, e.g., x^* was obtained by rounding $\bar{x} \in F^*(\bar{P}_1)$. So now we have a finite cutting value $z^* = cx^* < \infty$. Let now $v(\bar{P}_3) > z^*$, so P_3 is fathomed by (4.4). We have now P_2, P_4 and P_5 on the list of subproblems. For further analysis, we may choose any of them. Let us assume that it is P_4 and let P_4 be branched into P_6 and P_7.

If $\bar{x} \in F^*(\bar{P}_6)$ and $\bar{x} \in F(P)$, i.e., the optimal solution of the linear programming relaxation is an integer for all $j \in N_c$, then P_6 is fathomed by (4.5), and we have a new incumbent $x^* = \bar{x}$ and $z^* = c\bar{x}$. If $v(\bar{P}_7) \geqslant z^*$, then P_7 is fathomed. Now the list of subproblems consists of P_2 and P_5. We may choose any of them. Any choice corresponds to the so-called *backtracking* as both P_2 and P_5 are not successors of a currently considered vertex P_7 and they were constructed before P_7. We stop the computations when the list of subproblems is empty.

4.1.2. *Remarks about the Efficiency of a Branch-and-Bound Method*

Results of computational experiments and theoretical considerations show that, in discussing the efficiency of any branch-and-bound method, we have to answer the following three questions.

(1) How to efficiently branch a given problem into subproblems? To simplify the considerations, let us consider branching into only two subproblems. By the way, this type of branching, due to simplicity of its organization, is commonly used. Assume that $\bar{x} \in F^*(\bar{P})$ and $\bar{x}_r$ is not an integer for some $r \in N_c$. Then PP_1 is obtained from P by adding one constraint

$$x_r \leqslant \lfloor \bar{x}_r \rfloor, \tag{4.7}$$

and PP_2 differs from P by the constraint

$$x_r \geqslant \lfloor \bar{x}_r \rfloor + 1. \tag{4.8}$$

The variable x_r is called the *branching variable*. Usually, there are many variables $\bar{x}_j$ which are not integers for $j \in N_c$. Then we are asking which one of them should be chosen as the branching variable.

(2) In the previous subsection, we showed how linear programming can be used in the calculation of bounds. Obviously, the tighter the bound, more subproblems are fathomed by (4.4) and (4.5). Results of computational experiments show that the linear programming bounds are, in general, quite weak estimations of $v(PP_i)$ and one may ask how to efficiently find sufficiently good bounds.

(3) Usually, there are many subproblems on the list of subproblems, and then it is natural to ask which subproblem should be chosen for the analysis.

Answering these questions is not easy. It is quite possible that, e.g., the best, in a certain sense, choice of a branching variable will require much more time than solving a given problem with some quite simple and definitely not best choices of a branching variable. The same is true for the case of calculation of bounds.

It is interesting to note that the answer to these questions has changed with time. At the beginning of the branch-and-bound methods, in the sixties, it was commonly considered that it is better to analyse more subproblems than calculate tighter bounds. Later on, in the seventies, the converse statement became more popular: it is useful to calculate as tight as possible bounds on $v(PP_i)$ to reduce the number of subproblems. As a result, there are many methods for improving bounds. We will consider some of them in the next section and in Chapter 7.

Usually, the better bound allows to fathom more subproblems by (4.4). Recent computational experiments show that it is worth reformulating a given problem in such a way that a new formulation does not differ very much from its linear programming relaxation. Then there are good chances that many problems may be fathomed by (4.5). Such equivalent formulations will be considered in Chapter 6. In such reformations, we will often use the results for the knapsack problem considered in Chapter 5.

4.2. IMPROVING BOUNDS

It follows from Section 4.1 that branching, bounding and analysis of the problem P and any subproblem PP_i are the same. Therefore, we will further consider problem P only.

Usually, $v(\overline{P})$ gives a weak lower bound on $v(P)$ and this bound may be improved by adding the so-called *penalties* for violating integrality requirements for some x_r, $r \in N_c$. A penalty is any real number $k > 0$ such that

$$v(P) \geqslant v(\overline{P}) + k. \tag{4.9}$$

In this section, we show how to calculate better bounds using the simplex algorithm. For simplicity of presentation, we assume $N_c = N$ (Exercise 4.6.5).

Let $\bar{x} \in F^*(\overline{P})$ and assume that $\bar{x}_r$, $r \in N_c$, is not an integer. Consider $v(\overline{PP_i})$ as a function of x_r. A typical picture of such a function is given in Fig. 4.2 (Exercise 4.6.4). Since x_r must be an integer, we have

$$v(P) \geqslant \min \{v(\overline{P}|x_r \leqslant \lfloor \bar{x}_r \rfloor), v(\overline{P}|x_r \geqslant \lfloor \bar{x}_r \rfloor + 1)\}. \tag{4.10}$$

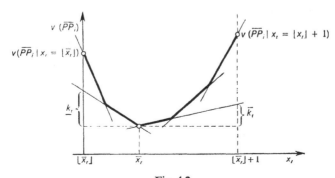

Fig. 4.2.

The values of arguments of the operator min in (4.10) may be estimated as $\underline{k}_r$ and $\bar{k}_r$ (see Fig. 4.2). We call $\underline{k}_r$ a *down penalty* and $\bar{k}_r$ *up penalty*. We show below that $\underline{k}_r$ and $\bar{k}_r$ may be computed in one simplex iteration. As now all variables must be integers ($N = N_c$), we have

$$k = \max_{r \in N} \min \{\underline{k}_r, \bar{k}_r\}. \tag{4.11}$$

Let

$$x_{Bi} = h_{i0} - \sum_{j \in R} h_{ij} x_j, \quad i = 0, 1, \ldots, m, \tag{4.12}$$

be an optimal simplex tableau for $\bar{P}$, where the zero row corresponds to the objective function, x_{Bi} is the basic variable, h_{i0} is its optimal value and R is the index set of nonbasic variables.

Assume that x_{Br} is not an integer. Then $x_{Br} = \lfloor h_{r0} \rfloor + f_{r0}$ with f_{r0} is the fractional part of x_{Br}. We know from Section 3.5 that the constraint (4.7) $x_{Br} \leqslant \lfloor \bar{x}_{Br} \rfloor$ may be written as

$$s_r = -f_{r0} + \sum_{j \in R} h_{rj} x_j. \tag{4.13}$$

This constraint violates the primal feasibility of the simplex tableau (4.12) which remains dual feasible. So we may apply the dual simplex algorithm (Section 2.7). The variable entering the basis is

$$\frac{h_{0k}}{h_{rk}} = \min_{j \in R} \left\{ \frac{h_{0j}}{h_{rj}} \,\middle|\, h_{rj} > 0 \right\}.$$

Then after one simplex iteration, the value of the objective function increases by $f_{r0} h_{0k}/h_{rk}$. Therefore, the down penalty $\underline{k}_r$ may be estimated as

$$\underline{k}_r = \min_{j \in R} \{f_{r0} h_{0j}/h_{rj} | h_{rj} > 0\}. \tag{4.14}$$

If in (4.14) all $h_{rj} \leqslant 0$, then (4.7) is redundant in $\bar{P}$, and we may fathom the subproblem corresponding to constraint (4.7). Therefore, we make $\underline{k}_r = \infty$.

Consider now constraint (4.8). It may be written in the form

$$s = f_{r0} - 1 - \sum_{j \in R} h_{rj} x_j. \tag{4.15}$$

In a way similar as in (4.14), we obtain the up penalty

$$\bar{k}_r = \min_{j \in R} \{(f_{r0} - 1) h_{0j}/h_{rj} | h_{rj} < 0\}. \tag{4.16}$$

If all $h_{rj} \geqslant 0$, constraint (4.8) is redundant in $\bar{P}$ and we may fathom the subproblem corresponding to (4.8). Then $\bar{k}_r = \infty$.

Knowing $\underline{k}_r$ and $\bar{k}_r$ for all $r \in N$ we may compute the penalty k by (4.11) and improve the bound on $v(P)$.

The requirement for x_j, $j \in N$, to be an integer means that some nonbasic variables have to equal at least one, which gives us a new penalty

$$k' = \min_{j \in R} h_{0j}. \tag{4.17}$$

Another penalty is available from the cut obtained by the method of integer forms from the rth row of the simplex tableau (4.12)

$$s_r = -f_{r0} + \sum_{j \in R} f_{rj} x_j$$

(Section 3.5). We may use this constraint to obtain a penalty k'' as an increase of h_{00} after one iteration of the dual simplex algorithm:

$$k'' = \min \{h_{10} h_{0j}/f_{rj} | f_{rj} \neq 0\}.$$

So now we may compute the lower bound on $v(P)$ as

$$z = v(\bar{P}) + \max \{k, k,', k''\}. \tag{4.18}$$

4.3. THE CHOICE OF A SUBPROBLEM

If the list of subproblems contains more than one subproblem, it is natural to ask which one of them should be chosen for analysis. In this section, we consider the rules most commonly used for the choice of a subproblem from a given list.

One of the most commonly used rules is called the *LIFO* (Last In–First Out) *rule.* According to this rule, we choose among the two new created subproblems the one for which the lower bound

$$\underline{z}_i = v(\overline{PP_i}) + \max \{k, k', k''\}$$

(compare (4.18)) is smaller. The organization of the list of subproblems is simple and, additionally, if PP_1 and PP_2 are obtained from P, an optimal tableau for $\bar{P}$ is dual feasible for both $\overline{PP}_1$ and $\overline{PP}_2$. One can check that using this rule we always consider only a part of the tree of subproblems, e.g., the left part of the tree. Therefore, the rule is sometimes called the *partial choice rule.*

The opposite rule to the LIFO rule is the *full choice rule* in which we are choosing a subproblem with the smallest lower bound z_i over all subproblems on the list. This requires a more sophisticated data structure. Often in practice the so-called *mixed choice rule* is used which can be formulated in the following way.

Let z^+ and z^- be the maximal and minimal values of lower bounds for all subproblems on the list, and let z' be the smallest lower bound among two newly created subproblems if they both are added to the list, or the lower bound of the subproblem added to the list as the last, e.g., if the other subproblem is fathomed by (4.5). Let $0 \leqslant \alpha \leqslant 1$. If

$$z' > \alpha z^+ + (1 - \alpha) z^-, \tag{4.19}$$

then, according to the mixed choice rule, we select the subproblem corresponding to z^-, otherwise we select the subproblem corresponding to z'. It is easy to see that for $\alpha = 0$, (4.19) becomes the rule of full choice, and for $\alpha = 1$, we have from (4.19) the LIFO rule. Relatively good results are obtained for $\alpha = 0.8$ and, in general, it is suggested to change α in the course of computations.

So far in choosing PP_i we are taking into account only an estimation of $v(PP_i)$. It follows from computational experiments that better results are obtained if we additionally take into account the violation of integer requirements. A simple measure of such a violation is defined for $\bar{x} \in F^*(\overline{PP_i})$ in the following way:

$$s_i = \sum_{j \in N_c} \min\{f_j, 1-f_j\}. \tag{4.20}$$

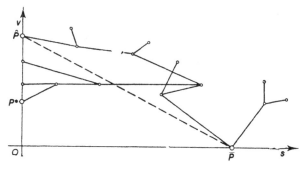

Fig. 4.3.

Then each subproblem PP_i may be represented as a point (s_i, v_i) on a plane with $v_i = v(PP_i) - v(\overline{P})$ (see Fig. 4.3). The linear programming relaxation $\overline{P}$ is lying on the line $v = 0$. By $\hat{P}$ we denote the first subproblem for which (4.5) holds, and by P^* we denote the subproblem which gives an optimal solution to a given problem P. Obviously, P^* is the lowest located point on the line $s = 0$. In Fig. 4.3 we join the subproblems (points) PP_i and PP_j if PP_j is obtained from PP_i by adding one of constraints (4.7) or (4.8), or vice versa.

Usually, we have no points (subproblems) below the line $\overline{P}P^*$ (see Fig. 4.3) and the best choice would be a selection of a subproblem which is close to the line $\overline{P}P^*$. As we do not know P^* before the end of computations, we use the line $\overline{P}\hat{P}$ as an approximation of $\overline{P}P^*$ (we asume that $\hat{P}$ has appeared so far) and consider points (subproblems) in the triangle $\overline{P}O\hat{P}$.

Dealing with points on a plane may be cumbersome, therefore we project points (s_i, v_i) on the line $s = 0$ parallel with the line $\overline{P}\hat{P}$, which gives a new lower bound on $v(PP_i)$ (compare 4.18):

$$z_i = v(\overline{PP_i}) + \beta s_i = v(\overline{PP_i}) + \sum_{j \in N_c} \min\{\beta f_j, \beta(1-f_j)\}, \tag{4.21}$$

where β is a tangent of the line $\overline{P}\hat{P}$.

We may interpret β in (4.21) as a cost of changing an integer variable by one. Better results are obtained if we allow different costs $\underline{\beta}$ and $\bar{\beta}$ for changing an integer variable down and up, respectively. Then these values are called *pseudocosts* and are defined in the following way:

$$\underline{\beta}_j = \frac{v(\overline{PP_1}) - v(\overline{P})}{f_j} \quad \text{and} \quad \bar{\beta}_j = \frac{v(\overline{PP_2}) - v(\overline{P})}{1 - f_j}, \tag{4.22}$$

where, as usual, we assume that PP_1 and PP_2 are obtained from P by adding (4.7) and (4.8), respectively. Since in large problems x_j may be a branching variable many times, it is suggested to take $\underline{\beta}_j$ and $\bar{\beta}_j$ as an average of values computed in each such branching by (4.22).

If PP_1 and PP_2 are obtained from P by branching along x_r, we may use simpler estimations than (4.22):

$$z_1 = v(\bar{P}) + \underline{\beta}_r f_r + \sum_{j \in N_c - \{r\}} \min\{\underline{\beta}_j f_j, \bar{\beta}_j (1-f_j)\}, \tag{4.23}$$

$$z_2 = v(\bar{P}) + \bar{\beta}_r (1-f_r) + \sum_{j \in N_c - \{r\}} \min\{\underline{\beta}_j f_j, \bar{\beta}_j (1-f_j)\}. \tag{4.24}$$

These estimations are simpler since we do not use $v(\overline{PP_i})$ in them.

Having a new incumbent x^* we usually check the first criteria of fathoming (4.4) for all subproblems on the list. Computational experiments show that it is useful to compute the so-called *percentage error* defined in the following way:

$$p_i = \frac{cx^* - w_i}{cx^* - z_i} \times 100, \tag{4.25}$$

where w_i is a pseudocost estimation on $v(PP_i)$ computed by (4.23) or (4.24), while z_i is computed by (4.18). If there are large negative errors, this implies that a better integer solution is very likely to be found. Therefore, we select a subproblem PP_i with the smallest percentage error p_i.

4.4. THE SELECTION OF A BRANCHING VARIABLE

In the selection of a branching variable, we have the same objectives as in the selection of a subproblem from the list. Usually, there are many variables which may be chosen as a branching variable. We select a variable with the highest priority, where we call a *priority of a variable* x_j, $j \in N_c$, a number $\gamma_j \geqslant 0$ that defines the influence of the choice of x_j as a branching variable on the efficiency of a branch-and-bound method.

In general, we have *static and dynamic lists of priorities*. In the first case, the priorities of variables do not change in the course of computations and only the set $N_c(i) = \{j \in N_* | \bar{x} = \lfloor \bar{x}_j \rfloor + f_j, f_j > 0\}$ changes. As usual, $\bar{x} \in F^*(\overline{PP_i})$. The priorities should reflect the "importance" of the integer variable to which they are attached; the more important the variable, the higher the priority. The importance may be specified in several ways. Perhaps the most satisfactory way of determining priorities is for the user to specify them from his knowledge of the physical problem. The other way is the experience gained from solving problems similar to the given one.

In the case of dynamic priorities, γ_j and $N_c(i)$ change in each iteration. They may be determined and changed basing on lower bounds or pseudocosts (Exercise 4.6.7).

In many problems, we have constraints of the form

$$\sum_{j \in Q} x_j = 1, \quad x_j = 0 \text{ or } 1, \quad j \in Q. \tag{4.26}$$

The set Q is called a *special ordered set*. They appear in the multiple-choice knapsack problems which we will discuss in the next chapter and in nonconvex separable problems (see also Section 1.5).

Let $\bar{x}$ be an optimal solution to the linear programming relaxation of P in which we have constraints (4.26) and let $0 < \bar{x}_r < 1$. The typical branching using (4.7) and (4.8) gives $x_r = 0$ and $x_r = 1$. The PP_1 is almost the same as P, while in PP_2 there is no constraint (4.26). In other words, PP_2 has a different structure than P.

More efficient is partitioning of Q in two almost equal subsets Q_1 and Q_2 in the following way. From $1 < \bar{x}_r < 1$ follows the existence of at least one index $r' \in Q$ such that $0 < \bar{x}'_r < 1$. Then we choose Q_1 and Q_2 in such a way that $r \in Q_1, r' \in Q_2$, $Q_1 \cup Q_2 = Q$, $Q_1 \cap Q_2 = \varnothing$ and $|Q_1| \approx |Q_2|$. Thus, instead of (4.7) and (4.8), we have

$$\sum_{j \in Q_1} x_j = 1 \quad \text{and} \quad \sum_{j \in Q_2} x_j = 1 \tag{4.27}$$

(Exercise 4.6.8).

4.5. BIBLIOGRAPHIC NOTES

Geoffrion and Marsten (1972) is a basic reference on branch-and-bound methods. Later works by Mitra (1973), Breu and Burdet (1974), Forrest et al. (1974), Crowder et al. (1983) contain results of computational experiments showing big progress in the efficiency of the branch-and-bound method. Theoretical aspects of this approach are considered in Mitten (1970). Description of branch-and-bound methods are discussed in all of the books on integer programming given in Section 1.7. The idea of penalties was introduced by Driebeek in 1966 and developed in the above-mentioned works.

4.6. EXERCISES

4.6.1. Formulate a branch-and-bound algorithm for the case of binary variables.

4.6.2. Solve the problem

$$\begin{aligned}
v(P) = \max(2x_1 + \ &x_2 + 4x_3 + \ 5x_4), \\
x_1 + \ 3x_2 + 2x_3 + \ &5x_4 \leqslant \ 10, \\
2x_1 + 16x_2 + \ x_3 + \ &x_4 \geqslant \ 4, \\
3x_1 - \ x_2 - 5x_3 + &10x_4 \leqslant \ -4, \\
x \geqslant 0, \quad x &\in Z^4,
\end{aligned}$$

using the algorithm described in Section 4.1. How is the tree of subproblems changed if we use the penalties for violations of integer requirements?

4.6.3. Give the condition for the finiteness of the branch-and-bound without assumption about the boundness of $F(P)$.

4.6.4. Consider $v(PP_i)$ as a function of x_r. Prove that it is convex and piecewise linear.

4.6.5. Give a method for calculating penalties for mixed integer programming problems.

4.6.6. Estimate penalties in the case of bounded variables.

4.6.7. Derive priorities from the pseudocosts (Forrest et al., 1974).

4.6.8. Give a method for calculating penalties for special ordered sets.

4.6.9. Assume that rules for branching and bounding are fixed. Prove that then the full choice rule gives the smallest number of subproblems to be considered.

CHAPTER 5

The Knapsack Problem

The binary knapsack is one of the most important problems in discrete programming. It has many practical applications, discussed in Section 1.5, and often appears as a subproblem in the analysis and solving of more complicated problems. Although it is $\mathcal{NP}$-hard, i.e., the difficulty of solving the knapsack problem is, generally speaking, the same as the difficulty of solving a general integer programming problem, the most advanced results in solving the linear programming relaxation, analysis of branch-and-bound and dynamic programming methods and description of the convex hull of feasible solutions have been obtained for the knapsack problem.

The aim of this chapter is to describe the most important results concerning the exact methods for solving knapsack problem. Near-optimal methods will be considered in Chapter 9. In Section 5.6, we consider the so-called multiple-choice knapsack problem which is a generalization of the binary knapsack problem.

In this chapter, we will consider mainly the binary knapsack problem only and ask in the exercises to generalize some of these results to the case of the integer knapsack problem. For simplicity, we will not repeat adjective binary.

5.1. RELAXATIONS

The knapsack problem may be formulated as

$$(K) \qquad v(K) = \max \sum_{j=1}^{n} c_j x_j \tag{5.1}$$

subject to

$$\sum_{j=1}^{n} a_j x_j \leqslant b, \tag{5.2}$$

$$x_j = 0 \text{ or } 1, \quad j \in N = \{1, \dots, n\}.$$

Now we show that, without loss of generality, we may assume that all data in K are positive integers.

If $c_j \leqslant 0$ and $a_j \leqslant 0$, after substitution $x'_j = 1 - x_j$, we have all coefficients in (5.1) and (5.2) nonnegative, with the appropriate change of b. If $c_j > 0$ and $a_j \leqslant 0$, then $x_j^* = 1$ in any optimal solution $x^* \in F^*(K)$. Similarly, if $c_j \leqslant 0$ and $a_j > 0$, then $x_j^* = 0$ in any $x^* \in F^*(K)$. So we may assume that all a_j, c_j are positive integers. If now $b < 0$, then K is infeasible. So the necessary condition for the knapsack problem to be feasible is $b \geqslant 0$. If $a_j > b$, then $x_j^* = 0$ in any $x^* \in F^*(K)$. Thus, without loss of generality, we may assume that all data in K are integers and

$$c_j > 0, \quad 0 < a_j \leqslant b, \quad \text{and} \quad \sum_{j=1}^{n} a_j > b. \tag{5.3}$$

If the last condition in (5.3) does not hold, K has a trivial solution $x^* = (1, 1, \ldots, 1)$.

Although K has a very simple structure, i.e., linear objective function and one linear constraint, it is an $\mathcal{NP}$-hard problem and thus branch-and-bound methods should be expected to be an efficient approach to solving K. Therefore, we are interested in efficient methods for solving the linear and Lagrangean relaxations of K, which we will study in the next two subsections.

5.1.1. *The Linear Programming Relaxation*

The linear programming relaxation of K, sometimes called the *linear knapsack problem*, is defined as

$$v(\bar{K}) = \max \left\{ \sum_{j=1}^{n} c_j x_j \, \bigg| \, \sum_{j=1}^{n} a_j x_j \leqslant b, 0 \leqslant x_j \leqslant 1, j \in N \right\}. \tag{$\bar{K}$}$$

The dual problem to $(\bar{K})$ takes then the form

$$v(\bar{D}) = \min \left(bu + \sum_{j=1}^{n} w_j \right) \tag{$\bar{D}$}$$

subject to

$$a_j u + w_j \geqslant c_j, \quad j = 1, \ldots, n,$$
$$u \geqslant 0, \quad w_j \geqslant 0, \quad j = 1, \ldots, n$$

Let $\bar{x}$ be an optimal solution to $\bar{K}$, i.e., $\bar{x} \in F^*(\bar{K})$, and let $(\bar{u}, \bar{w}) \in F^*(\bar{D})$. Since $\bar{K}$ is a linear programming problem with one constraint and n bounded variables, from Section 2.5, by (5.3), we know that there exists one basic variable $0 < \bar{x}_p \leqslant 1$. From the complementarity slackness (Theorem 2.7), we have that $\bar{x}$ and $(\bar{u}, \bar{w})$ have to satisfy the following system of linear equalities:

$$\bar{x}_j(a_j \bar{u} + \bar{w}_j - c_j) = 0, \quad j = 1, \ldots, n, \tag{5.4}$$
$$\bar{w}_j(1 - \bar{x}_j) = 0, \quad j = 1, \ldots, n, \tag{5.5}$$
$$\bar{u} \left(b - \sum_{j=1}^{n} a_j \bar{x}j \right) = 0. \tag{5.6}$$

Then, for $j = p$, from (5.4) and (5.5) we have

$$\bar{u} = c_p / a_p. \tag{5.7}$$

The constraints of $\bar{D}$ may be written as

$$w_j \geqslant c_j - a_j u \quad \text{and} \quad w_j \geqslant 0 \quad \text{for} \quad j = 1, \ldots, n.$$

Since we have minimization in $\bar{D}$, we have

$$w_j = \max\{0, c_j - a_j u\} \quad \text{for} \quad j = 1, \ldots, n.$$

We call the ratio $r_j = c_j/a_j$ the *efficiency* of the jth variable, $j \in N$, and, without loss of generality, we may assume

$$r_1 \geqslant r_2 \geqslant \ldots \geqslant r_n. \tag{5.8}$$

Now we show that two vectors

$$\bar{x} = \left(1, \ldots, 1, \left(b - \sum_{j=1}^{p-1} a_j\right) \Big/ a_p, 0, \ldots, 0\right), \tag{5.9}$$

$$(\bar{u}, \bar{w}) = (r_p, c_1 - a_1 r_p, \ldots, c_{p-1} - a_{p-1} r_p, 0, \ldots, 0) \tag{5.10}$$

are optimal solutions to $\bar{K}$ and $\bar{D}$, respectively. Obviously, $\bar{x}$ is a feasible solution to $\bar{K}$ and $(\bar{u}, \bar{w}) \in F(\bar{D})$. Since (5.9) and (5.10) satisfy the complementarity slackness conditions (5.4)–(5.6), we have

THEOREM 5.1. *An optimal solution to the linear programming relaxation of the knapsack problem satisfying (5.3) and (5.8) is given by (5.9). The optimal solution to its dual is given by (5.10).* □

We note that the index p always exists due to (5.3), i.e., $1 \leqslant p \leqslant n$. The value $\bar{u} = c_p/a_p = r_p$ is called *critical efficiency*.

COROLLARY 5.1. *The ordering (5.8) is not, in general, unique, and the optimal solution (5.9) is neither unique, while the threshold value $r_p = c_p/a_p$ is unique as is the optimal solution (5.10) to the dual problem $\bar{D}$.* □

Thus to solve $\bar{K}$, we have to order variables according to (5.8), which requires $O(n \log n)$ operations, and next to compute $\bar{x}$ by (5.9), which may be done in $O(n)$ operations. Such an approach is called the primal method as it in fact solves the primal linear programming problem $\bar{K}$. In the next subsection, we show that it is possible to solve the dual $\bar{D}$ without sorting, i.e., in $O(n)$ operations. Thus, by the duality theory, it is possible to solve $\bar{K}$ in linear time.

5.1.2. A Linear-Time Algorithm

We may express $v(\bar{P})$ as a function of a single variable u. Since $w_j = \max\{0, c_j - a_j u\}$, $j \in N$, we have

$$v(\bar{D}) = \min f(u) = \min\left(bu + \sum_{j=1}^{n} \max\{0, c_j - a_j u\}\right).$$

Observe that $f(u)$ is a convex piecewise linear function with the breaking points $r_j = c_j/a_j$, $j \in N$.

Thut to solve $\bar{D}$, it is enough to find a critical efficiency $\bar{u} = r_p$ minimizing $f(u)$.

Note that such a $\bar{u}$ exists as the cases $\bar{u} = 0$ and $\bar{u} = \infty$ are ruled out by (5.3). To find the optimal point $\bar{u}$, we compute for a given u the left-hand derivative $\partial^- f(u)$ and the right-hand derivative $\partial^+ f(u)$ given as

$$\partial^- f(u) = b - \sum_{j | r_j \geq u} a_j \quad \text{and} \quad \partial^+ f(u) = b - \sum_{j | r_j > u} a_j.$$

As a consequence, to solve $\bar{D}$ and at the same time $\bar{K}$, it is enough to find the following partition of the set $N = N_1 \cup \{p\} \cup N_0$ such that

$$r_j \geq r_p \geq r_q \quad \text{for each} \quad j \in N_1, q \in N_0,$$

$$\sum_{j \in N_1} a_j < b \leq \sum_{j \in N_1} a_j + a_p.$$

Then the optimal solution $\bar{x} \in F^*(\bar{K})$ is given by

$$\bar{x}_j = 1 \quad \text{for } j \in N_1, \quad \bar{x}_p = \left(b - \sum_{j \in N_1} a_j\right) \Big/ a_p, \quad \bar{x}_j = 0 \quad \text{for } j \in N_0.$$

In other words, we have to solve two inequalities

$$S_1(r) < b \leq S_2(r), \tag{5.11}$$

where

$$S_1(r) = \sum_{j | r_j > r} a_j \quad \text{and} \quad S_2(r) = S_1(r) + \sum_{j | r_j = r} a_j.$$

We put $S_1 = 0$ if $\{j | r_j > r\}$ is empty.

The algorithm below solves (5.11) without sorting $r_j, j \in N$. For a given x, we define $N_0 = \{j | x_j = 0\}$, $N_1 = \{j | x_j = 1\}$ and $N_f = N - (N_1 \cup N_0)$, so N_f is an index set of free, not fixed, variables. We will always assume that $S_1(r)$ and $S_2(r)$ are computed for $j \in N_f$.

Algorithm A1 (A Linear-Time Algorithm for $\bar{K}$)
Step 0 (Initialization): $N_0 = \emptyset$; $N_1 := \emptyset$; $N_f := N$; $\bar{b} := b$;
Step 1 (Selection of r): Select $r = r_j$, $j \in N_f$; Partition N_f into three subsets
$\quad N^+ := \{j \in N_f | r_j > r\}$; $N^= := \{j \in N_f | r_j = r\}$; $N^- := \{j \in N_f | r_j < r\}$;
$\quad$ Compute $S_1(r)$ and $S_2(r)$;
Step 2 (Optimality test):
$\quad$ (a) If $S_1(r) < \bar{b} \leq S_2(r)$ (r is right), then stop;
$\quad$ (b) If $S_1(r) \geq \bar{b}$ (r is too small), then $N_0 := N_0 \cup N^- \cup N^=$; $\quad N_f := N^+$;
$\quad\quad$ go to Step 1;
$\quad$ (c) If $S_2(r) < \bar{b}$ (r is too large), then $N_1 := N_1 \cup N^+ \cup N^=$; $\quad N_f := N^-$;
$\quad\quad \bar{b} := \bar{b} - S_2(r)$; go to Step 1. $\qquad\qquad\qquad\qquad\qquad\qquad \square$

In Step 2(a) we have the solution $x_j = 1$ for $j \in N_1 \cup N^+$ and $x_j = 0$ for $j \in N_0 \cup N^-$, and $\bar{x}_j$ for $j \in N^=$ is computed by the so-called filling the knapsack procedure, in which we in turn put $\bar{x}_j = 1$; $\bar{b} := \bar{b} - a_j$ until $a_j < \bar{b}$. At the end, we have $\bar{x}_j = \bar{b}/a_j \leq 1$ and $\bar{x}_k = 0$ for the remaining $k \in N^=$.

Now we estimate the computational complexity of Algorithm A1. The median of the set $\{r_j | j \in N\}$ is the $\lceil n/2 \rceil$th element of this set. For instance, for $n = 7$, the median is the fourth element. Finding the median requires less operations than sorting a given set. It may be shown (see Section 5.7 for references) that it requires not more than $3n + O((n \log n)^{3/4})$ comparisons. Choosing r as the median of $\{r_j | j \in N_f\}$ in Step 1 of Algorithm A1 guarantees that at least half of the variables are eliminated from N_f in each iteration. In the worst case, Algorithm A1 stops when $|N_f| \leqslant n/2^k \leqslant 1$, where k is the number of iterations. Thus $k \leqslant \lceil \log n \rceil$.

At the beginning, we make n divisions computing $r_j = c_j/a_j$, $j \in N$. At Step 1, we make at most $3|N_f|$ comparisons to find the median of N_f (we may drop $O((n \log n)^{3/4})$ in our estimation), $|N_f|$ comparisons to partition $|N_f|$ into N^+, $N^=$ and N^-, and $|N_f|$ summations to compute $S_1(r)$ and $S_2(r)$. Thus we make $5|N_f|$ operations. At Step 2, we make 2 comparisons and change 6 pointers of the beginnings and ends of the sets. Thus at Step 2, we make 8 operations.

Let t_1 be an execution time for division and t_2 be an execution time for comparison, summation and changing the pointers. By $T(n)$ we denote the maximal time of solving $\bar{K}$ by Algorithm A1. Then

$$T(n) \leqslant t_1 n + t_2 5n(1 + 1/2 + 1/4 + \ldots + 1/2^k) + 8t_2 k + C(k)$$
$$\leqslant (t_1 + 10t_2)n + 8t_2 \log n,$$

where $C(k)$ grows not faster than $O(k)$. Therefore $T(n) = O(n)$. So we have proved

THEOREM 5.2. *If r is chosen as the median of $\{r_j | j \in N_f\}$, then Algorithm A1 has the complexity $O(n)$.* $\qquad\qquad\square$

Computational experiments show that for small sets N_f, say with $|N_f| \leqslant 20$, it is better to order N_f according to (5.8). In practice, the median is computed in a near-optimal way, e.g., we take as a median of N_f, the median of the first three elements of N_f.

Algorithm A1 is the *dual method* for $\bar{K}$ as in fact it solves $\bar{D}$. Since $2n + 1$ data are needed to formulate $\bar{K}$, Algorithm A1 is an optimal algorithm within a constant factor.

5.1.3. *Lagrangean Relaxations*

For a given $\lambda \geqslant 0$, the Lagrangean relaxation of K is an integer programming problem of the form

$$L(\lambda) = \max(cx + \lambda(b - ax)) \qquad\qquad (L)$$

subject to

$$x \in \{0, 1\}^n.$$

So in L_λ we are looking for a maximum of a linear function on the vertices of the unit cube in R^n. Obviously, L is a relaxation of K as $F(K) \subseteq F(L_\lambda)$ and $cx + \lambda(b - ax) \geqslant cx$ for any $x \in F(K)$ and any $\lambda \geqslant 0$ (Section 1.3).

In general, we say that the Lagrangean relaxation has the *integrality property* if and only if its value does not change if the integrality requirements are dropped. It is easy to see that L_λ has the integrality property, i.e., $L(\lambda)$ does not change if we substitute the condition $x_j = 0$ or 1 by $0 \leqslant x_j \leqslant 1$, $j \in N$. So we may write $L(\lambda)$ as

$$L(\lambda) = b\lambda + \max_{0 \leqslant x_j \leqslant 1} \sum_{j=1}^{n} (c_j - a_j \lambda) x_j = b\lambda + \sum_{j=1}^{n} \max\{0, c_j - a_j \lambda\}.$$

Observe that the above formulation gives the value of the objective function in the dual problem $\overline{D}$ for the feasible solution $u = \lambda \geqslant 0$ and $w_j = \max\{0, c_j - a_j \lambda\}$, $j \in N$. From Section 1.3 we know that

$$v(\overline{K}) \geqslant L(\lambda) \geqslant v(K),$$

and from the integrality property of L_λ follows that the best possible choice of λ is

$$\lambda^* = \bar{u} = r_p = c_p/a_p,$$

and then $L(\lambda^*) = v(\overline{K})$. So we have proved

THEOREM 5.3. *For* $\lambda^* = c_p/a_p = \bar{u}$, *we have*

$$L(\lambda^*) = v(\overline{K}) = v(\overline{D}).\qquad\qquad\qquad\square$$

So in the case of K, the Lagrangean relaxation does not give better bounds than the linear programming. Observe that $L(\lambda)$ may be computed in $O(n)$ for a given $\lambda \geqslant 0$, namely, we put $x_j = 1$ if $c_j - a_j \lambda > 0$, $x_j = 0$ for $c_j - a_j \lambda < 0$ and arbitrarily $x_j = 0$ or 1 if $c_j - a_j \lambda = 0$.

5.2. REDUCTION METHODS

The aim of the reduction of the number of variables in K is to fix as many variables at their optimal values as possible before solving K. If $v(P)$ is the optimal value of a general integer programming problem P, then $\underline{v}(P)$ ($\overline{v}(P)$) is a lower (upper) bound on $v(P)$.

The general reduction scheme for the 0–1 problem is

if $\overline{v}(P|x_j = e_j) < \underline{v}(P)$, *then* $x'_j = 1 - e_j$ *in every solution* x' *of* P *such that* cx' $> \underline{v}(P)$.

By $P|x_j = e_j$, with $e_j = 0$ or 1, we denote the problem P with an additional constraint $x_j = e_j$. The power of such reduction depends on how tight the lower and upper bounds on $v(P)$ are.

In the case of the knapsack problem K, a relatively good lower bound can be computed by the so-called *greedy algorithm*: If $\bar{x} \in F^*(\overline{K})$, then at the beginning we put $x'_j = 1$ for $j \in N_1 = \{j | \bar{x}_j = 1\}$, and next, for each $j \in N - N_1$, we put $x'_j = 1$ if

$$a_j \leqslant b - \sum_{j \in N_1} a_j,$$

and update $N_1 := N_1 \cup \{j\}$. The greedy algorithm gives good results if (5.8) holds. In Chapter 9, we will estimate the maximal error of $x' \in F(K)$ obtained by the greedy algorithm.

The commonly used two reduction methods differ in the way of computing the upper bounds for $e_j = 1 - x'_j$, $j \in N$, where x' is the best feasible solution to K known so far.

(1) $\bar{v}(K|x_j = e_j) = \lfloor v(\bar{K}|x_j = e_j) \rfloor$, i.e., the upper bounds are obtained by the linear programming relaxation of K. In particular, if

$$v(\bar{K}|x_j = 0) < cx', \tag{5.12}$$

then in any feasible solution x better than x' we have $x' = 1$. Therefore, $x^* = 1$ in any optimal solution $x^* \in F^*(K)$. Then we reduce the number of variables in K by one. Similarly, if

$$v(\bar{K}|x_j = 1) < cx', \tag{5.13}$$

then $x_j^* = 0$.

(2) $\bar{v}(K|x_j = e_j) = \lfloor L(\lambda^*)|x_j = e_j \rfloor$, where $L(\lambda^*)$ is the value of the Lagrangean function for $\lambda^* = c_p/a_p$. Since $L(\lambda^*) = v(\bar{K})$, then $v(L(\lambda^*)|x_j = 1 - \bar{x}_j) = v(L(\lambda^*)) - |\bar{c}_j|$, where $\bar{x} \in F^*(\bar{K})$ and $\bar{c}_j = c_j - a_j \lambda^*$, $j \in N$. Therefore, the general reduction scheme gives

$$\text{if} \quad \lfloor L(\lambda^*) - |\bar{c}_j| \rfloor < cx', \quad \text{then} \quad x'_j = \operatorname{sgn} \bar{c}_j, \tag{5.14}$$

where $\operatorname{sgn} a = 1(0)$ if $a \geqslant 0$ ($a < 0$).

The first method requires solving n linear knapsack problems, therefore its complexity is $O(n^2)$, although some savings in computations are possible if we use the solution for a given j to obtain an optimal solution of $\bar{K}|x_q = e_q$ for $q \neq j$. The second method has a complexity $O(n)$, but is weaker, i.e., reduces not more but, in generally, fewer variables than the first method.

In practice, different reduction methods may be mixed and the following concept of *variable dominance* may be included. We say that x_i dominates x_j if and only if

$$c_i \geqslant c_j \quad \text{and} \quad a_i \leqslant a_j \tag{5.15}$$

and at least one inequality is strict. The obvious property of the dominance is: if x_i dominates x_j, then $x_i^* \geqslant x_j^*$. Thus if x_i dominates x_j, then we have the following corollary: if $i \in N_0$, then $j \in N_0$ and, on the other hand, if $j \in N_1$, then $i \in N_1$. Computational experiments show that including the dominance of variables in the reduction method increases its efficiency.

5.3. THE BRANCH-AND-BOUND APPROACH

In this section, we present general suggestions how to solve K by a branch-and-bound method described in Chapter 4. The efficiency of such an approach comes from the following facts.

First, it is possible to solve efficiently the linear knapsack problem in $O(n)$ by the dual method or in $O(n\log n)$ by the primal method.

Second, computational experiments show that it is possible to reduce, in general, 70–80% of n using the reduction tests from Section 5.2.

Third, it is possible to find a relatively good feasible solution to K, e.g., by the greedy algorithm in $O(n)$ or by near optimal methods described in Chapter 9.

First the linear programming relaxation $\bar{K}$ is solved by the dual method. If $\bar{x}_p \neq 1$, the reduction of the number of variables is done. Let x^* be the best feasible solution found so far. If $v(\bar{K}) - cx^* < 1$, then x^* is, obviously, an optimal solution. As in $\bar{x}$ we have only one fractional variable, the problem of choosing the branching variable is solved. Moreover, computational experiments suggest that instead of keeping the list of subproblems it is better to enumerate implicitly all binary vectors, starting from x^* until, in the worst case, $(0, 0, \ldots, 0)$ is reached. In this implicit enumeration, we assign variable x_j the priority $\gamma_j = 1/r_j, j \in N$.

To simplify description of the implicit enumeration, let us assume for the moment that (5.8) holds. Let x_k^* be the last one in x^*. We put $x_k^* = 0$ and solve the linear knapsack problem for variables $x_j, k+1 \leqslant j \leqslant n$. If $v(\bar{K}|x_k = 0) > cx^*$, we solve the corresponding subproblem in the same way as K. Otherwise, if $v(\bar{K}|x_k = 0) \leqslant cx^*$, we do the backtrack, i.e., we find the last one in x^*, let it be $x_i^* = 1$. Next we compute $v(\bar{K}|x_i = 0)$ and so on. Since $i < k-1$, after a finite number of iterations, we reach in the worst case $(0, 0, \ldots 0)$. A reduction method may also be applied to each subproblem.

5.4. DYNAMIC PROGRAMMING

In this section, we will consider an integer knapsack problem

$$v(K) = \max \sum_{j=1}^{n} c_j x_j,$$

$$\sum_{j=1}^{n} a_j x_j \leqslant b,$$

$$x_j \geqslant 0 \quad \text{and integer}, j = 1, \ldots, n.$$

Obviously, the binary knapsack problem is a special case of the integer one. Without loss of generality, we may assume that all data in K are positive integers.

Let $v_k(y)$ be the value of the knapsack subproblem defined for the first k variables and for the right-hand side $b = y$

$$v_k(y) = \max \left\{ \sum_{j=1}^{k} c_j x_j \,\middle|\, \sum_{j=1}^{k} a_j x \leqslant y, \, x_j \geqslant 0, \, x_j \in Z, \, j = 1, \ldots, k \right\}. \tag{5.16}$$

If $k \geqslant 2$, then for $y = 0, 1, \ldots, b$, we may write (5.16) in the form

$$v_k(y) = \max_{x_k = 0, 1, \ldots, \lfloor y/a_k \rfloor} c_k x_k +$$

$$+ \max \left\{ \sum_{j=1}^{k-1} c_j x_j \,\middle|\, \sum_{j=1}^{k-1} a_j x_j \leqslant y - a_k x_k, \, x_j \geqslant 0, \, x_j \in Z, \, j = 1, \ldots, k-1 \right\}.$$

The expression in the brackets equals $v_{k-1}(y - a_k x_k)$. So we may write (5.16) as

$$v_k(y) = \max_{x_k = 0, 1, \ldots, \lfloor y/a_k \rfloor} \{c_k x_k + v_{k-1}(y - a_k x_k)\}. \tag{5.17}$$

Putting $v_0(y) = 0$ for $y = 0, 1, ..., b$ we extend (5.17) for the case $k = 1$. The relation (5.17) expresses the so-called *dynamic programming principle of optimality* which says that regardless of the number of the kth item chosen, the remaining space, $y - a_k x_k$, must be allocated optimally over the first $k-1$ items. In other words, looking for an optimal decision at the n-stage process, we have to take an optimal decision at each stage of the process.

Observe that (5.17) was obtained without any assumption about the constraint and the objective function. In fact, the dynamic programming principle may be easily generalized to treat problems with a nonlinear objective function and/or constraint, and also problems with bounded variables.

If, for a given y and k, there is an optimal solution to (5.17) with $x_k = 0$, then $v_k(y) = v_{k-1}(y)$. On the other hand, if $x_k > 0$, then in an optimal solution to (5.17) one item of the kth type is combined with an optimal knapsack of size $y - a_k$ over the first k items. Thus we have

$$v_k(y) = \max \{v_{k-1}(y), c_k + v_k(y - a_k)\} \tag{5.18}$$

for $k = 1, ..., n$ and $y = 0, 1, ..., b$. Then obviously $v(Z) = v_n(b)$. If the knapsack constraint changed to an equality, (5.18) still applies with the initial condition $v_0(0) = 0$, and $v_0(y) = -\infty$ otherwise.

The computation of $v_k(y)$ requires by (5.18) comparison of two numbers. Thus the computational complexity of the dynamic programming is $O(nb)$. Dynamic programming is not a polynomial algorithm for solving K, since the length of data in K is $O(n \log n)$.

Computational experiments show that the time for solution of a given knapsack problem by dynamic programming does not depend on the ordering of variables. Thus, if within T' we solved by dynamic programming a problem with n' variables, the solution time for the problem with $n \gg n'$ variables may be estimated as $T = T'n/n'$. We use this observation in a hybrid computer code KNAPSOL which combines dynamic programming with branch-and-bound (see Section 5.7 for references). First T is estimated in the above way for a given binary knapsack problem. If the problem is not solved by the branch-and-bound method within T, the computations are automatically switched to the dynamic programming method.

5.5. THE KNAPSACK POLYTOPE

The convex hull of the 0–1 solution to the single inequality $ax \leqslant b$ is called the *knapsack polytope*, so

$$W = \text{conv}\{x \in R^n | ax \leqslant b, x \in \{0, 1\}^n\} = \text{conv} F(K).$$

The *dimension* of W, $\dim W$, is the smallest dimension of the real space containing W. By (5.3), $\dim W = n$ and $W \neq \emptyset$. An equality $hx = h_0$ in R^n is a *supporting hyperplane* of W if $hx \leqslant h_0$ for all $x \in W$ and $H = W \cap \{x \in R^n | hx = h_0\} \neq \emptyset$. The polytope H and the corresponding inequality $hx \leqslant h_0$ are called a *face* of W. If $\dim H = n-1$, then H is called a *facet* of W. If $\dim H = 1$, then it is called an *edge* of W, and if dim $H = 0$, then W is called a *vertex* of W (see Fig. 5.1). A complete list of facets of W is still unknown to date and some facet inequality may be obtained from a minimal cover defined below.

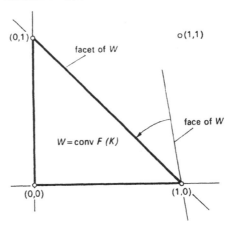

Fig. 5.1.

A subset S of N is called a *minimal cover* of $ax \leqslant b$ if

$$\sum_{j \in S} a_j > b \quad \text{and} \quad \sum_{j \in S} a_j - a_k \leqslant b \quad \text{for all} \quad k \in S. \tag{5.19}$$

Every $x \in F(K)$ has to satisfy

$$\sum_{j \in S} x_j \leqslant |S| - 1. \tag{5.20}$$

Usually, $|S| < n$ and (5.20) defines only a face of W. To obtain a facet, i.e., the best possible inequality of W, (5.20) must be "lifted" by the so-called *lifting procedure*: initially, we set $h = 1$ for $j \in S$ and $h_0 = |S| - 1$. Next, for $k \in N - S$, we compute

$$z_k = \max \left\{ \sum_{j \in S} h_j x_j \,\middle|\, \sum_{j \in S} h_j x_j \leqslant b - a_k, \ x_j = 0 \ \text{or} \ 1, \ j \in S \right\},$$

define $h_k = h_0 - z_k$, redefine S as $S \cup \{k\}$ and repeat until $N - S = \emptyset$.

THEOREM 5.4. *The constraint* $hx \leqslant h_0$ *obtained by the lifting procedure is a facet of* W. $\qquad\qquad\square$

5.6. THE MULTIPLE-CHOICE KNAPSACK PROBLEM

An interesting generalization of the knapsack problem is obtained if we add constraints on the choice of variables. If the set of variables N is partitioned into m classes N_k, $k = 1, \ldots, m$, and if we require that exactly one variable has to be chosen from each class, we get the *multiple-choice knapsack* problem (MK), which may be formulated as

$$v(MK) = \max \sum_{k=1}^{m} \sum_{j \in N_k} c_j x_j \tag{MK}$$

subject to

$$\sum_{k=1}^{m} \sum_{j \in N_k} a_j x_j \leqslant b,$$

$$\sum_{j \in N_k} x_j = 1, \quad k \in M = \{1, \ldots, m\},$$

$$x_j = 0 \text{ or } 1, \quad j \in N = \bigcup_{k=1}^{m} N_k = \{1, \ldots, n\}.$$

Such a problem has similarly as K many applications, e.g., in capital budgeting, menu planning and in solving nonlinear knapsack problems.

Without loss of generality, we may suppose that all data are integers and, to avoid trivial situations, we assume that

$$\sum_{k=1}^{m} \min_{j \in N_k} a_j \leqslant b < \sum_{k=1}^{m} \max_{j \in N_k} a_j. \tag{5.21}$$

The problem is $\mathcal{N}\mathcal{P}$-hard as K is its special case. To see this, add n slack variables in constraints $x_j \leqslant 1$, $j \in N$, in K. In this way, we obtain the equivalent MK with $m = n$ and $|N_k| = 2$ for all $k \in M$.

The problem is also interesting from a theoretical point of view. First, it often appears as a part of a general integer programming problem (see Chapter 7). Second, since we know (Section 5.1) that it is possible to solve $\bar{K}$ in $O(n)$, it is natural to ask about a possibility of solving the linear programming relaxation of the multiple-choice knapsack problem in linear time. In the next subsection, we give a positive answer to this question.

5.6.1. Relaxations of MK

The linear programming relaxation of MK, sometimes called the *linear multiple-choice knapsack problem* and denoted as $\overline{MK}$, is obtained by replacing $x_j = 0$ or 1 by $x_j \geqslant 0$, $j \in N$. We may drop constraints $x_j \leqslant 1$, $j \in N$, because of the multiple-choice constraints

$$\sum_{j \in N_k} x_j = 1, \quad k \in M.$$

This is a linear programming problem with $m+1$ constraints and may be solved by the simplex method, but due to the special structure of the constraints, it can be done in a more efficient way. To do this, we study the dual to $\overline{MK}$, which may be formulated as

$$v(\overline{MD}) = \min\left(bu + \sum_{k=1}^{m} w_k\right) \tag{$\overline{MD}$}$$

subject to

$$a_j u + w_k \geqslant c_j \quad \text{for} \quad j \in N_k, \; k \in M,$$
$$u \geqslant 0.$$

In a way similar as in Section 5.1, it is easy to see that $\overline{MD}$ is equivalent to minimization of a convex piecewise linear function

$$v(\overline{MD}) = f(u) = \min_{u \geqslant 0} \left(bu + \sum_{k=1}^{m} \max_{j \in N_k} (c_j - a_j u) \right)$$

$$= \min_{u \geqslant 0} \left(bu + \sum_{k=1}^{m} f_k(u) \right), \qquad (5.22)$$

where $f_k(u) = \max_{j \in N_k} (c_j - a_j u)$ for $k \in M$. Figure 5.2 shows how $f(u)$ is constructed in the case $k = 2$ and $|N_1| = |N_2| = 3$.

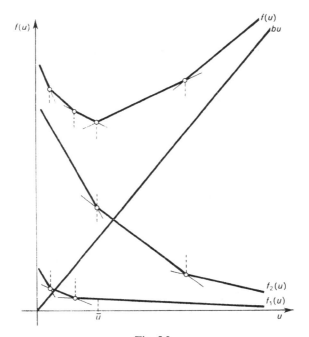

Fig. 5.2.

Thus, to find $(\bar{u}, \bar{w}) \in F^*(\overline{MD})$, it is enough to find the minimum $\bar{u}$ of the convex piecewise linear function of a single variable u, $f(u)$. Obviously, $f(u)$ attains its minimum at $\bar{u}$ if and only if

$$\partial^- f(\bar{u}) \leqslant 0 \leqslant \partial^+ f(\bar{u}), \qquad (5.23)$$

where $\partial^- f(u)$ and $\partial^+ f(u)$ are the left-hand derivative and the right-hand derivative of the function f at point u, respectively. By (5.22), they are equal

$$\partial^- f(u) = b - \sum_{k=1}^{m} \max_{j \in N_k} \{a_j | c_j - a_j u = f_k(u)\}, \qquad (5.24)$$

$$\partial^+ f(u) = b - \sum_{k=1}^{m} \min_{j \in N_k} \{a_j | c_j - a_j u = f_k(u)\}. \qquad (5.25)$$

Since $N_i \cap N_j = \emptyset$ for $i \neq j$, we have either two or no fractional variables in an optimal solution $\bar{x}$ to $\overline{MK}$, and if $\bar{x}$ has two fractional variables, they are in the same class N_p, $1 \leqslant p \leqslant m$, say $j'_p, j'_p \in N_p$.

Consider now a vector in R^n defined as

$$\bar{x}_j = \begin{cases} 1 & \text{for } j = j_k, k \neq p, \text{ where } f_k(\bar{u}) = c_{j_k} - a_{j_k}\bar{u}, \\ \bar{b}/(a_{j_p} - a_{j_p}) & \text{for } j = j'_{p'}, \\ 1 - \bar{x}_{j'_p} & \text{for } j = j_p, \\ 0 & \text{otherwise}, \end{cases} \tag{5.26}$$

where

$$\bar{b} = b - \sum_{k=1}^{m} a_{j_k} \geqslant 0.$$

Now we construct a $(m+1)$-dimensional vector $(\bar{u}, \bar{w}) = (\bar{u}, \bar{w}_1, \ldots, \bar{w}_m)$ with

$$\bar{u} = (c_{j'_{p'}} - c_{j_p})/(a_{j'_p} - a_{j_p}) \quad \text{and} \quad \bar{w}_k = (c_{j_k} - a_{j_k}\bar{u}), \quad k \in M. \tag{5.27}$$

It is not difficult to show that $\bar{x} \in F(\overline{MK})$, $(\bar{u}, \bar{w}) \in F(\overline{DM})$ and they both satisfy the complementarity slackness conditions. Thus we have

THEOREM 5.5. $\bar{x} \in F^*(\overline{MK})$ and $(\bar{u}, \bar{w}) \in F^*(\overline{DM})$.   □

So any optimal solution to $\overline{MK}$ has at most two fractional variables $0 \leqslant x_{j'_p} < 1$ and $0 < x_{j_p} \leqslant 1$, $j'_p, j_p \in N_p$, $1 \leqslant p \leqslant m$. An optimal solution $\bar{x}$ is, in general, not unique, but $(\bar{u}, \bar{w})$ is unique.

As there are

$$\frac{1}{2} \sum_{k=1}^{m} |N_k|^2 = O(n^2)$$

breaking points, the straightforward method of computing (5.24) and (5.25) at these breaking points has the complexity $O(n^2)$. Now we show that it can be done in $O(n)$.

To do this, we group lines (variables) $c_j - a_j u$, $j \in N_k$, in $\lfloor |N_k|/2 \rfloor$ pairs. Then, for a given u and a given pair $c_i - a_i u$ and $c_j - a_j u$, we define an intersection point (if any) $u_{ij} = (c_i - c_j)/(a_i - a_j)$, $a_i \neq a_j$, $i, j, \in N_k$, $k \in M$, and consider three possible cases.

(a) If $c_j \geqslant c_i$ and $a_j = a_i$, then the lines are parallel and $c_i - a_i u$ is *dominated* by $c_j - a_j u$, in the sense of (5.22). Therefore, $c_i - a_i u$ may be eliminated from N_k. See case (a) in Fig. 5.3.

(b) If $c_j \geqslant c_i$ and $a_j < a_i$, then $u_{ij} < 0$, and since we are looking for $\bar{u} > 0$, the line $c_i - a_i u$ may be eliminated from N_k (case (b) in Fig. 5.3). Observe that the cases $\bar{u} = 0$ and $\bar{u} = \infty$ are eliminated by (5.3) and (5.21).

(c) If $c_j \geqslant c_i$ and $a_j > a_i$, then

$$c_j - a_j u \geqslant c_i - a_i u \quad \text{for any } u \leqslant u_{ij},$$
$$c_i - a_i u \geqslant c_j - a_j u \quad \text{for any } u \geqslant u_{ij}.$$

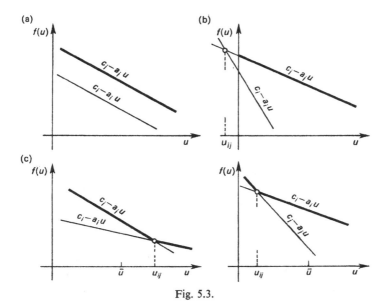

Fig. 5.3.

Then, if $\partial^- f(u_{ij}) > 0$, the line $c_i - a_i u$ may be eliminated. On the other hand, if $\partial^+ f(u_{ij}) < 0$, the line $c_j - a_j u$ may be eliminated (see cases (c) in Fig. 5.3).

So, in any case, one element of a pair may be eliminated. This elimination constitutes the essence of the *dual method* for $\overline{MK}$. We assume that all elements of N_k, $k \in M$, are somehow grouped in pairs, and if lines i and j form a pair, we say that i is *adjacent* to j, and vice versa. Obviously, if $c_i - a_i u$ is dominated in the pair (i, j), then $x_i = 0$ in any $\bar{x} \in F^*(\overline{MK})$, and x_i may be eliminated from further considerations.

Algorithm A2

Step 0 (Initialization): $\overline{M} := M$; $\bar{b} := b$; $\overline{N}_k := N_k$ for $k \in M$;

Step 1 (Selection of u): Delete lines dominated according (a) and (b) in $\overline{N}_k$, $k \in \overline{M}$; If $\overline{N}_k = \{j_k\}$, then begin $\bar{b} := \bar{b} - a_{j_k}$; $\overline{M} := \overline{M} - \{k\}$ end; Choose $k \in \overline{M}$; Select $j \in \overline{N}_k$; $i :=$ adjacent to j in $\overline{N}_k$; $u := (c_i - c_j)/(a_i - a_j)$; $LD := \partial^- f(u)$; $RD := \partial^+ f(u)$;

Step 2 (Optimality test):

(1) If $LD \leqslant 0 \leqslant RD$, then $\bar{u} := u$ stop;

(2) If $LD > 0$, then for all $k \in \overline{M}$ and for all pairs in $\overline{N}_k$ do (if $(c_i - c_j)/(a_i - a_j) \geqslant u$, then delete line according to (c)); Go to Step 1;

(3) If $RD < 0$, then for all $k \in \overline{M}$ and for all pairs in $\overline{N}_k$ do (if $(c_i - c_j)/(a_i - a_j) \leqslant u$, then delete line according to (c)); Go to Step 1.   □

If at Step 1 we select u in the following way:

$$u = \text{med}\{(c_p - c_q)/(a_p - a_q)|(p, q) \text{ is a pair in } \overline{N}_k, k \in \overline{M}\}, \tag{5.28}$$

i.e., u is selected as the median of all current pairs, then we have

THEOREM 5.6. *The dual method with (5.28) solves $\overline{MK}$ in linear time.*

Proof. As in each iteration there are

$$\sum_{k\in\overline{M}} \lfloor |\overline{N}_k|/2 \rfloor$$

of pairs, at least

$$\left[\left(\sum_{k\in\overline{M}} \lfloor |\overline{N}_k|/2 \rfloor \right) / 2 \right] \geq \left(\sum_{k\in\overline{M}} |\overline{N}_k| \right) / 6$$

lines are eliminated. Then the total complexity of the dual method is bounded by

$$O\left(\sum_{i=1}^{\infty} (5/6)^i |\overline{N}_i| \right) = O(|N|) = O(n). \qquad\qquad \square$$

One may easily find similarities between Algorithm A1 and Algorithm A2. In the next subsection, we show how the complexity of the dual method can be improved using dominance relations among variables.

5.6.2. Dominance Relations

We say that j *dominates* i if $i, j \in N_k$, $k \in M$, $c_j \geq c_i$ and $a_j \leq a_i$.

THEOREM 5.7. *If j dominates i, there exists $x^* \in F^*(MK)$ with $x_i^* = 0$.*

Proof. Let $x' \in F^*(MK)$ with $x_i' = 1$. We construct the other vector x^* with $x_i^* = 0$, $x_j^* = 1$ and $x_l^* = x_l'$ for all other $l \in N$. This vector is feasible as $ax^* = ax' + a_j - a_i \leq ax' \leq b$ and, moreover, $cx^* = cx' + c_j - c_i \geq cx'$. So either $x^* \in F^*(MK)$ or we get a contradiction. $\qquad\qquad \square$

The problem with undominated items is called the *undominated multiple-choice knapsack problem* (undominated MK). Any algorithm for computing the set of undominated items has the complexity

$$O\left(\sum_{k=1}^{m} |N_k| \log |N_k| \right) = O(n \log n)$$

as it requires sorting of items in each class according to c_j or a_j and next deleting the dominated items, which can be done in $O(n)$.

So now we may assume that besides (5.3) for the undominated MK, all classes N_k are reindexed such that if $N_k = \{p_k, \ldots, q_k\}$, $p_k < q_k$, then

$$\begin{aligned} 0 \leq a_{p_k} &< a_{p_k+1} < \ldots < a_{q_k}, \\ 0 \leq c_{p_k} &< c_{p_k+1} < \ldots < c_{q_k}. \end{aligned} \tag{5.29}$$

For an undominated MK with (5.29), we say that i, j *linear dominate* h if $p_k \leq i < h < j \leq q_k$, $k \in M$, and $(c_h - c_i)/(a_h - a_i) \leq (c_j - c_h)/(a_j - a_h)$.

THEOREM 5.8. *If i, j linear dominate h, there exists $\bar{x} \in F^*(\overline{MK})$ with $\bar{x}_h = 0$.*

Proof. Let $x' \in F^*(\overline{MK})$ with $x'_h \neq 0$. Similarly as in the proof of the previous theorem, we construct a vector $\bar{x}$ with $\bar{x}_h = 0$, $\bar{x}_i = x'_i + x'_h(a_j - a_h)/(a_j - a_i)$, $\bar{x}_j = x'_j + x'_h(a_h - a_i)/(a_j - a_i)$, and $\bar{x}_l = x'_l$ for all $l \in N$. Next we show that $\bar{x} \in F(\overline{MK})$ and moreover $c\bar{x} > cx'$. So either $\bar{x} \in F^*(\overline{MK})$ or we get a contradiction. □

The problem with only linear-undominated items is called the *linear-undominated MK problem*. Due to (5.29), it may be computed in linear time. Again without loss of generality, we may assume that now $N_k = \{r_k, \ldots, s_k\}$, $r_k < s_k$, $k \in M$, and

$$0 \leqslant a_{r_k} < a_{r_k+1} < \ldots < a_{s_k},$$
$$0 \leqslant c_{r_k} < c_{r_k+1} < \ldots < c_{s_k}, \tag{5.30}$$
$$0 > d_{r_k+1} > d_{r_k+2} > \ldots > d_{s_k},$$

where $d_j = (c_j - c_{j-1})/(a_j - a_{j-1})$ and $r_k + 1 \leqslant j \leqslant s_k$. Figure 5.4 shows the relations between different points (a_j, c_j). Observe that points satisfying (5.30) form a piecewise linear concave function.

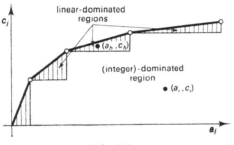

Fig. 5.4.

Now, due to (5.30), the algorithm for finding the median in ordered sets (see Section 5.7 for references) can be applied and the complexity of the dual method can be further decreased. To see this, observe that the overfilling item j'_p is adjacent to j_p in N_p, i.e., $j'_p = j_p + 1$, thus

$$v(\overline{MK}) = \min_{\substack{i \in N_k \\ i \neq r_k}} \left\{ bd_i + \sum_{k=1}^{m} \max_{j \in N_k} \{c_j - a_j d_i\} \right\}$$

and $\bar{u} = d_{j_p+1}$. Now we modify (5.28) in the following way:

$$u = \text{med}\{d_j | j \in \bar{N}_k, j \neq r_k, k \in \bar{M}\}. \tag{5.31}$$

THEOREM 5.9. *The dual method for the linear-undominated $\overline{MK}$ with (5.30) has the complexity $O(m \log^2(n/m))$.* □

We will not prove the above theorem here (see Section 5.7 for the references) and only note that the dual method for the linear-undominated $\overline{MK}$ is very efficient in the case when n/m is large and in fact this method is sublinear in n. It is important that sorting should be done only once during solving $\overline{MK}$.

5.7. Bibliographic Notes

Dudziński and Walukiewicz (1987) reviewed exact methods for the knapsack problem and its generalizations. They pointed out the importance of the dual methods for solving linear programming relaxations.

5.1. Dantzig (1957) was the first who pointed out the importance of the knapsack problem in integer programming and gave the expression (5.9) for $\bar{x} \in F^*(\bar{K})$. The computational complexity of sorting is estimated in the book by Aho et al. (1974). The optimal procedure for finding the median is given in Schönhage et al. (1976). Algorithm A1 was proposed, among others, by Johnson and Mizogushi (1978), Balas and Zemel (1980), Fayard and Plateau (1982).

5.2. Tests (5.12) and (5.13) are given by Ingargiola and Korsh (1973). Walukiewicz (1975) proposed to use the dominance relations in these tests. Dembo and Hammer (1980) and Nauss (1976) used the Lagrangean relaxation in reduction methods. The case of the integer knapsack problem was studied by Ingargiola and Korsh (1977).

5.3. Many efficient branch-and-bound algorithms have been proposed for the binary knapsack problem, e.g., by Fayard and Plateau (1982), Martello and Toth (1978), Balas and Zemel (1980), Zoltners (1978), Walukiewicz (1975). These papers contain results of computational experiments. From the computational results we conclude that a knapsack problem with randomly generated coefficients and n up to 10 000 may be solved on modern computers in less than a second of CPU time. Algorithm A1 plays a very important role in these codes as the ordering time constitutes, generally speaking, the major part of the total solution time.

5.4. Dynamic programming was applied to solving the knapsack problem by Bellman (1957) and Dantzig (1957). Gilmore and Gomory (1961, 1963, 1965, 1966) used dynamic programming in the so-called cutting stock problem. A description of KNAPSOL is given in Walukiewicz (1975).

5.5. Balas (1975) and Hammer et al. (1975) described methods for constructing some facets of the knapsack polytope. The lifting procedure is given in Balas and Zemel (1977).

5.6. Glover and Klingman (1979), developing the work of Witzgall (1979), gave an $O(n\log n)$ primal algorithm for solving $\overline{MK}$ and asked whether it is possible to solve it in $O(n)$. The positive answer to this question was given by Dyer (1984) who proposed Algorithm A2. The other algorithm for $\overline{MK}$ and MK were proposed by Ibaraki et al. (1978), Sinha and Zoltners (1979) and Zemel (1980). The modification of the dual method for the linear-undominated $\overline{MK}$ is given in Dudziński and Walukiewicz (1984). The computational complexity of finding the median in ordered sets is estimated in Frederickson and Johnson (1982). Some generalizations of the multiple-choice knapsack problem is given in Johnson and Padberg (1981) and in Armstrong et al. (1982).

5.8. EXERCISES

5.8.1. For the integer knapsack problem, modify (5.3) and give an expression for $\bar{x} \in F^*(\bar{K})$.

5.8.2. Solve by Algorithm A1

$$v(\bar{K}) = \max(19x_1 + 3x_2 + 8x_3 + x_4 + 2x_5 + 10x_6 + 4x_7 + 9x_8 + x_9 +$$
$$+ 7x_{10} + 19x_{11} + 18x_{12}),$$
$$2x_1 + 5x_2 + 7x_3 + 6x_4 + 3x_5 + 2x_6 + 15x_7 + 12x_8 + 8x_9 + 7x_{10} + 15x_{11} + x_{12}$$
$$\leqslant 18,$$
$$0 \leqslant x_j \leqslant 1, \quad j = 1, \ldots, 12.$$

5.8.3. Solve the problem from Exercise 5.8.2 by ordering the variables according to (5.8).

5.8.4. Reduce the number of variables in the problem from Exercise 5.8.2, assuming $x_j = 0$ or 1 for $j = 1, \ldots, 12$ using
   (a) relations (5.12) and (5.13),
   (b) relation (5.14),
   (c) relations (5.12) and (5.13) and the dominance relations.

5.8.5. Solve

$$v(K) = \max(11x_1 + 9x_2 + 5x_3 + 9x_4),$$
$$6x_1 + 5x_2 + 3x_3 + 12x_4 \leqslant 15,$$
$$x \geqslant 0, \quad x \in Z^4,$$

first by dynamic programming and next by the branch-and-bound method.

5.8.6. Prove Theorem 5.4.

5.8.7. Find all minimal covers for a given binary inequality

$$9x_1 + 7x_2 + 6x_3 + 4x_4 + 2x_5 + x_6 \leqslant 10.$$

Choose two of them and construct the corresponding facets of W.

5.8.8. Put $x_j = 0$ or 1, $j = 1, \ldots, 12$, in the problem of Exercise 5.8.2 and solve it using the reduction and branch-and-bound.

5.8.9. A problem

$$v(P) = \max \sum_{k=1}^{m} \sum_{j \in N_k} c_j x_j,$$

$$\sum_{k=1}^{m} \sum_{j \in N_k} a_j x_j \leqslant b,$$

$$\sum_{j \in N_k} x_j \leqslant h_k, \quad k = 1, \ldots, m,$$

with $N_i \cap N_k = \emptyset$ for $i \neq j$ and real data transform into an equivalent MK.

5.8.10. Transform a given MK into an equivalent knapsack problem (Zemel, 1980).

5.8.11. In Exercise 5.8.2 put $N_1 = \{1, 2, 3\}$, $N_2 = \{3, 4, 5\}$, $N_3 = \{6, 7, 8\}$, $N_4 = \{9, 10, 11, 12\}$ and solve $\overline{MK}$ by Algorithm A2.

5.8.12. Write a branch-and-bound algorithm for MK.

5.8.13. In Exercise 5.8.11 put $x_j = 0$ or 1, $j = 1, \ldots, 12$, and solve it by branch-and-bound method.

5.8.14. Write the greedy algorithm for MK.

5.8.15. Prove Theorem 5.9 (Dudziński and Walukiewicz, 1984).

CHAPTER 6

Equivalent Formulations for Integer Programs

In the first part of this chapter (Sections 6.1–6.3), we consider a method for constructing equivalent formulations for a following large integer programming problem

$$v(P) = \max\{cx|Ax \leqslant b,\ x \in \{0, 1\}^n\}.$$

The matrix A in such problems is usually sparse, i.e., the number of nonzero elements of A is a few per cent of the number mn, where m is the number of constraints of P. We also assume that A has no apparent special structure.

In Section 6.4, we will discuss Benders' decomposition (primal decomposition) in which a sequence of problems equivalent to a given mixed integer programming problem is constructed. Next, we describe the so-called primal-dual decomposition and, finally, in Section 6.6, we show that an integer programming problem is equivalent to the shortest path problem in an appropriately constructed graph.

6.1. THE REDUCTION OF THE SIZE OF BINARY PROBLEMS

The constraints of P can be classified into two types: type 1 constraints are special ordered set constraints and type 2 constraints are all of the other constraints of P.

A type 1 constraint may be written as

$$\sum_{j \in D} x_j - \sum_{j \in U} x_j \leqslant 1 - |U|, \tag{6.1}$$

where $D \cap U = \varnothing$ and $D \cup U \subseteq N = \{1, \ldots, n\}$. If $x_j = 1$ for some $j \in D$, then $x_k = 0$ for all $k \in D$, $k \neq j$, and $x_k = 1$ for all $k \in U$. Similarly, $x_j = 0$ for some $j \in U$ implies $x_k = 1$ for all $k \in U$, $k \neq j$, and $x_k = 0$ for all $k \in D$.

Consider now a type 2 constraint written for notational simplicity in the form

$$\sum_{j \in N_+} a_j x_j + \sum_{j \in N_-} a_j x_j \leqslant a_0, \tag{6.2}$$

where N_+ (N_-) is the index set of positive (negative) coefficients a_j. Clearly, if

$$\sum_{j \in N_-} a_j > a_0,$$

then no binary vector satisfies (6.2) and the problem P is infeasible. If

$$\sum_{j \in N_-} a_j \leqslant a_0,$$

then every binary vector satisfies (6.2), and such an inequality may be dropped from the constraint set of P, so we may reduce m by one in this case. If, for $j \in N_+$,

$$a_j > a_0 - \sum_{k \in N_-} a_k,$$

then $x_j = 0$ in any $x \in F(P)$. Therefore, we may reduce the number of variables in P by one and proceed to the analysis of the constraints of type 1. Likewise, if for some $j \in N_-$,

$$-a_j > a_0 - \sum_{k \in N_-} a_k,$$

then $x_j = 1$ in any $x \in F(P)$. Then we reduce n by one, adjust the right-hand-side vector b and go to the analysis of type 1 constraints. Further reduction of n is possible if in P we find a type 1 constraint and $j \in D$. Obviously, we consider type 2 constraints in the above way one by one as they appear in problem P.

6.2. THE ROTATION OF A CONSTRAINT

In Section 1.2, we define the equivalence between two integer problems and show that a given problem may have infinitely many equivalent formulations. The solution times for two equivalent formulations may differ very much providing they both are solved on the same computer and by the same method. Computer experiments indicate that the so-called *tighter equivalent formulation* is easier to solve.

Having two equivalent formulations P_1 and P_2 we say that the problem (formulation) P_2 is tighter than P_1 if $F(\bar{P_2}) \subseteq F(\bar{P_1})$, where, as usual, $\bar{P}$ is the linear programming relaxation of P. In this chapter, when we say that P_1 and P_2 are equivalent, we mean $F(P_1) = F(P_2)$. Thus, if P_1 and P_2 are binary problems, P_2 is tighter than P_1 if $F(P_2) = F(P_1)$ and $F(\bar{P_2})$ is a smaller part of the unit cube in R^n than $F(\bar{P_1})$.

Below we provide theoretical justification for such reasoning. Assume for a moment that P has only type 2 constraints. If we knew the convex hull of $F(P)$, i.e., if we knew all facets of the polytope

$$W = \text{conv} F(P) = \text{conv} \{x \in R^n | Ax \leqslant b, x_j = 0 \text{ or } 1, j \in N\},$$

then, instead of solving the integer problem P, it would suffice to solve the linear programming problem

$$v(P) = \max \{cx | x \in \text{conv} F(P)\}.$$

Unfortunately, finding all facets of W is a difficult problem and up to now we do not even know all of the facets of the knapsack polytope which, in general, is much simpler than W.

Let $a_i x \leqslant b_i$ be the ith constraint of P, $i = 1, \ldots, m$. This constraint corersponds to the knapsack polytope

$$W_i = \text{conv} \{x \in R^n | a_i x \leqslant b_i, x_j = 0 \text{ or } 1, j \in N\}.$$

It is easy to see that

$$W \subseteq \bigcap_{i=1}^m W_i. \tag{6.3}$$

Equality in (6.3) does not, in general, hold, but does hold if, for example, problem P decomposes totally into m knapsack problems. If we have a large-scale binary problem with a sparse matrix A and with no apparent special structure, it is reasonable to expect that the intersection of m knapsack polytopes W_i provides a fairly good approximation to W over which we wish to minimize a linear objective function.

The quality of such an approximation may be improved in at least two ways. First, we obtain a better description of W_i by rotating $a_i x \leqslant b_i$, $i = 1, \ldots, m$, which we will consider in this section. In the second way, we generate additional constraints which, together with $a_i x \leqslant b_i$, better describe W_i, $i = 1, \ldots, m$ (Section 6.3).

6.2.1. *The Rotation Procedure*

Consider a constraint of type 2 which, without loss of generality, may be written in the form

$$\sum_{j \in N} a_j x_j \leqslant a_0, \qquad x_j = 0 \text{ or } 1, \quad j \in N, \tag{6.4}$$

where $0 < a_j \leqslant a_0$ and a_j is an integer for all $j \in N$. Let F be the set of feasible solutions to (6.4).

We say that the *constraint* (cut) $\bar{a}x \leqslant \bar{a}_0$ is *stronger* than (6.4) if $\bar{a}_j \geqslant a_j, j \in N$, $\bar{a}_0 \leqslant a_0$ and at least one inequality is strict (see Section 3.7). If for any $r \in N$ there exists $x \in F$ such that

$$\sum_{j \in N-\{r\}} a_j x_j + a_r = a_0, \tag{6.5}$$

it is impossible to construct an inequality preserving conv F stronger than (6.4) and therefore (6.4) is called a *strongest constraint* (cut). Since $a_j > 0$ for all $j \in N$, we have $\bar{a}_0 = a_0$.

It is easy to see that in R^2 the construction of a strongest constraint is equivalent to rotation of $ax \leqslant a_0$, more precisely, it is equivalent to rotation of $ax = a_0$. Therefore, the method of construction of a strongest constraint is called the *rotation procedure*.

For all $r \in N$ and $x \in F$, we have

$$ax_r \leqslant a_0 - \sum_{j \in N-\{r\}} a_j x_j = a_0 - d_r.$$

Thus $a_r \leqslant a_0 - d_r$ for any $x \in F$ such that $x_r = 1$. If

$$d_r^* = \max\left\{ \sum_{j \in N-\{r\}} a_j x_j \,\Big|\, x \in F, x_r = 1 \right\},$$

then $a_r = a_0 - d_r^*$ for all $r \in N$. The value d_r^* may be computed as the maximal element of the set

$$D_r = \left\{ d \,\Big|\, d = \sum_{j \in J} a_j \quad \text{for all } J \subseteq N - \{r\},\ d \leqslant a_0 - a_r \right\}.$$

If during the computation of D_r for $r \geqslant 2$ we find that $d = a_0 - a_r$, then we put $\bar{a}_r = a_r$ and go to the computation of D_{r+1}.

Now we give a formal description of the rotation procedure.

Algorithm A3

Step 0 (Initialization, $r = 1$): $D_1^n = \{0, a_n\}$; For $i = n-1,\, n-2, ..., 1$ compute
  $D_1^i = D_1^{i+1} \cup \{d+a_i | d \in D_1^{i+1},\, d+a_i \leqslant a_0 - a_1\}$; Find d_1^*; $\bar{a}_1 := a_0 - d_1^*$;

Step 1 (Main iteration, $2 \leqslant r \leqslant n-1$): $r := r+1$; If $r = n$, go to Step 2; D_r^n
  $:= \{d | d \in D_1^{r+1},\, d \leqslant a_0 - a_r\}$; For $i = r-1, r-2, ..., 1$ compute $D_r^i := D_r^{i+1}$
  $\cup \{d+a_i | d \in D_r^{i+1},\, d+a_i \leqslant a_0 - a_r\}$; $D_r := D_r^1$; Find d_r^*; $\bar{a}_r := a_0 - d_r^*$;

Step 2 (Stop, $r = n$): $D_n^1 := \{0, a_1\}$; For $i = 2, 3, ..., n-1$, compute $D_n^i := D_n^{i-1}$
  $\cup \{d+\bar{a}_i | d \in D_n^{i-1},\, d+\bar{a}_i \leqslant a_0 - \bar{a}_n\}$; Find d_n^*; $a_n := \bar{a}_0 - d_n^*$; stop.   □

The rotation procedure gives the strongest constraint $\bar{a}x \leqslant a_0$ with $a_j \leqslant \bar{a}_j \leqslant a_0$. One can see that $ax \leqslant a_0$ is a strongest cut if it cannot be rotated, i.e., if $\bar{a}_j = a_j$ for all $j \in N$.

THEOREM 6.1. *If* $L = \{x \in F | ax = a_0\}$ *and* $\bar{L} = \{x \in F | \bar{a}x = a_0\}$, *then* $|\bar{L}| \geqslant |L|$, $L \subseteq \bar{L}$ *and* $|\bar{L}| > 2$. *The computational complexity of Algorithm A3 is* $O(n^2(a_0 - a_m))$, *where*

$$a_m = \min_{j \in N} a_j.$$ □

The value of $\bar{a}_j$ depends on the order in which this coefficient is computed by Algorithm A3. In chosing the order, it is practical to take into account the objective function, which together with $ax \leqslant a_0$ gives the knapsack problem. Let K be the knapsack problem with $ax \leqslant 0$, while K_0 the knapsack problem corresponding to the rotated constraint $\bar{a}x \leqslant a_0$. Since $F(\bar{K}_0) \subseteq F(\bar{K})$, it is reasonable to compute $\bar{a}_j$ in the same order as (5.8) in Section 5.1. In Exercises 6.8.5 and 6.8.6, we ask about the other methods for computing an equivalent constraint.

6.3. CONSTRAINT GENERATION

From Section 3.7 we know that the ideal additional constraint (cut) would be such a facet of W which cuts off $\bar{x} \in F^*(\bar{P})$ from $F(P)$. Up to now, we not know how to construct such a facet of W, but some progress is made if instead of W we consider the knapsack polytope W_i, $i = 1, ..., m$. The problem may be formulated more precisely in the following way.

For a given constraint of type 2

$$\sum_{j \in N} a_j x \leqslant a_0, \tag{6.6}$$

with all $a_j > 0$, find a minimal cover $S \subseteq N$ (see Section 5.5) such that the corresponding minimal cover inequality

$$\sum_{j \in S} x_j \leqslant |S| - 1 \tag{6.7}$$

cuts off $\bar{x}$ from $F(P)$, where $\bar{x} \in F^*(\bar{P})$. The above problem is called the *constraint identification problem*. Knowing S we may construct a facet of the knapsack polytope corresponding to (6.6) by the lifting procedure (Section 5.5).

Consider the following knapsack problem

$$q = \min \left\{ \sum_{j \in N} (1 - \bar{x}_j) y_j \;\Big|\; \sum_{j \in N} a_j y_j > a_0, \; y_j = 0 \text{ or } 1, \; j \in N \right\}, \tag{6.8}$$

where the knapsack constraint is a strict inequality and $\bar{x} \in F^*(\bar{P})$ is given.

THEOREM 6.2. *For a given constraint* (6.6), *there exists a minimal cover inequality* (6.7) *which cuts off x from* $F(P)$ *if and only if* $q < 1$.

Proof. ($\Rightarrow$) Let S be the minimal cover we are looking for, i.e., for a given $\bar{x} \in F^*(\bar{P})$ and for any binary x satisfying (6.2), we have

$$\sum_{j \in S} x_j \leqslant |S| - 1, \quad \text{but} \quad \sum_{j \in S} \bar{x}_j > |S| - 1.$$

Setting $y_j = 1$ for all $j \in S$ and $y_j = 0$ for all $j \in N - S$, we find that the knapsack constraint in (6.8) is satisfied and

$$q \leqslant \sum_{j \in S} (1 - \bar{x}_j) = |S| - \sum_{j \in S} \bar{x}_j < |S| - (|S| - 1) = 1.$$

($\Leftarrow$) If $q < 1$, then from $0 \leqslant \bar{x} \leqslant 1$ follows that $1 - \bar{x}_j \geqslant 0$. Let y^* be a unique optimal solution to (6.8). Define $S = \{j \in N | y_j^* = 1\}$. Then

$$1 > q = \sum_{j \in S} (1 - \bar{x}_j) = |S| - \sum_{j \in S} \bar{x}_j,$$

and (6.7) cuts off $\bar{x}$ from $F(P)$. Since y^* is a unique optimal solution to (6.8), S must be a minimal cover to (6.6). If (6.8) has many optimal solutions, there exists among them at least one minimal cover to (6.6). $\square$

Thus Theorem 6.2 solves the constraint identification problem providing in fact a most violated minimal cover inequality. We may extend it from S to N by the lifting procedure and obtain in this way a facet of the knapsack polytope corresponding to (6.6), i.e., the best possible cut for this polytope. The main advantage of such cuts, in comparison to cuts discussed in Chapter 3, is that they preserve the sparsity of A. For instance, if $hx \leqslant h_0$ is a facet of the polytope corresponding to (6.6), then $h_j \neq 0$ only if $a_j \neq 0$ in (6.6).

The facet $hx \leqslant h_0$ is added to the constraints of P and we solve the constraint identification problem for the next constraint of type 2. If for (6.6) there does not exist a minimal cover inequality (6.7) cutting off $\bar{x}$ from $F(P)$, we also consider the next constraint. Therefore, up to m additional constraints (cuts) may be added. Next we solve the linear programming relaxation of the problem with additional constraints. Let x' be an optimal solution of the relaxation. If x' is binary, then $x' \in F^*(P)$. If it is not the case, then we may solve the constraint identification problem with x'. In practice, we go to the construction of additional constraints if $cx - cx' \geqslant \varepsilon$, i.e., when x' gives a substantial improvement of $v(P)$, e.g., if $\varepsilon > 1$. If $cx - cx' < \varepsilon$, then we apply the branch-and-bound method. We do the same if we do not find a minimal cover inequality (6.7) cutting off $\bar{x}$ for any constraint of P.

6.3.1. *The Modified Lifting Procedure*

Solution of knapsack problems like (6.8) or those in the lifting procedure may be time consuming. Computational experiments indicate that a fairly good approximation of a facet $hx \leqslant h_0$ may be obtained after solving the above knapsack problems in a near-optimal way.

Consider the ith constraint of type 2 in P and let $\tilde{x} \in F^*(\overline{P})$. Using the substitution $x'_j = 1 - x_j$ where necessary, we bring the row under consideration into the form (6.6) with all nonnegative coefficients. This changes the solution vector from $\tilde{x}$ to, say, $\overline{x}$. Next we define

$$N_1 = \{j \in N \,|\, \overline{x}_j > 0,\, a_{ij} \neq 0\}, \quad N_0 = \{j \in N \,|\, x_j = 0,\, a_{ij} \neq 0\}.$$

Clearly, to solve the constraint identification problem, it is sufficient to work with the data in N_1 and usually $|N_1|$ is small for a problem with a sparse matrix. If $N_1 = \varnothing$, if $j \in N_1$ implies $\overline{x}_j = 1$ or if $j \in N_1$ implies $|a_{ij}| = 1$, then there is nothing to do in row i and we process the next type 2 constraint. Otherwise, we solve (6.8) with N replaced by N_1 as a linear programming problem, i.e., we replace $y_j = 0$ or 1 by $0 \leqslant y_j \leqslant 1$, $j \in N_1$. We know from Section 5.1 that at most one variable is fractional in an optimal solution $\overline{y}$ to the problem. The set $S_1 = \{j \in N_1 | \overline{y}_j > 0\}$ is a cover for (6.6) and may be easily changed into a minimal cover S by dropping some of the variables to zero.

Next we sort variables in S by decreasing the ratio $1/a_j$. If $N_1 - S \neq \varnothing$, we extend the minimal cover inequality (6.7) to the variables in $N_1 - S$ using the lifting procedure (Section 5.5). As in the previous case, we relax the zero-one problems solved in the lifting procedure

$$q_k = \max\left\{\sum_{j \in S} h_j x_j \,\Big|\, \sum_{j \in S} a_j x_j \leqslant a_0 - a_k,\, x_j = 0 \text{ or } 1,\, j \in S\right\} \tag{6.9}$$

to the corresponding linear knapsack problems by setting $0 \leqslant x_j \leqslant 1$, $j \in S$. Let $\overline{q}_k$ be the optimal value of this relaxation. Since $q_k < \lfloor \overline{q}_k \rfloor$, computing the coefficients for the facet $hx \leqslant h_0$ as $h_k = h_0 - \lfloor \overline{q}_k \rfloor$ we obtain the inequality $hx \leqslant h_0$, which is valid, i.e., it does not cut off any binary point x satisfying (6.6). It is a fairly good approximation of the facet of the corresponding knapsack polytope because $h_j \leqslant h_0 = |S| - 1$, and, in turn, $|S|$ is usually small in sparse problems. Ordering of variables in S usually increases the efficiency of the lifting procedure.

Such an approximation of a facet may not cut off $\overline{x}$ and we check whether

$$\sum_{j \in N_1} h_j \overline{x} > h_0. \tag{6.10}$$

If (6.10) does not hold, our attempt to construct a cut in a near-optimal way was unsuccessful and we proceed to the type 2 constraint. If (6.10) holds, we extend the corresponding inequality for $k \in N_0$ and add $hx \leqslant h_0$ to the constraints of P.

Thus, in Sections 6.2 and 6.3, we have described two methods of obtaining an equivalent formulation P_r of a given problem P. We have $F(P_r) = F(P)$, but

$F(\bar{P}_r) \subseteq F(P)$. The efficiency of such an approach may be measured as

$$e = \frac{v(\bar{P}) - v(\bar{P}_r)}{v(\bar{P}) - v(P)} . \tag{6.11}$$

Thus (6.11) measures the fraction of the duality gap closed due the constraint generation and/or the constraint rotation.

6.4. BENDERS' DECOMPOSITION

In this and the following sections, we will consider a mixed integer programming problem of the form

$$v(P) = \min(cx + dy) \tag{P}$$

subject to

$$Ax + Dy \geqslant b,$$
$$x, y \geqslant 0, \quad x \in Z^n,$$

where $d \in R^p$, $b \in R^m$ and matrices A and D have appropriate sizes.

For a fixed integer vector x, the problem reduces to the linear programming problem

$$v(Px) = cx + \min\{dy | Dy \geqslant b - Ax, y \geqslant 0\}. \tag{Px}$$

The dual to Px is

$$v(DPx) = cx + \max\{u(b - Ax) | uD \leqslant d, u \geqslant 0\}. \tag{DPx}$$

We observe that the convex set $F(DPx)$ does not depend on the choice of x. If $F(DPx) \neq \emptyset$, then $F(DPx)$ as the intersection of a finite number of half-spaces contains a finite number of vertices u^s, $s \in T_P$, and a finite number of extreme rays v^r, $r \in R_P$.

If the dual problem is inconsistent (see Section 2.6), i.e., if $F(DPx) = \emptyset$, then $F^*(P) = \emptyset$, as, in this case, either $v(Px) = -\infty$ or $F(Px) = \emptyset$. If $v(DPx) = +\infty$, the $F(Px) = \emptyset$ and as a consequence there is no feasible solution of P for that x. Thus x has to satisfy constraints

$$v^r(b - Ax) \leqslant 0 \quad \text{for all } r \in R_P.$$

So we may write DPx as

$$v(DPx) = cx + \max_{s \in T_p}\{u^s(b - Ax) | v^r(b - Ax) \leqslant 0, r \in R_p\}, \tag{DPx}$$

and the problem P in the form

$$v(P) = \min\left(cx + \max_{s \in T_p} u^s(b - Ax)\right),$$

subject to

$$v^r(b - Ax) \leqslant 0 \quad \text{for all } r \in R_P, \tag{6.12}$$
$$x \geqslant 0 \text{ and integer}.$$

Let

$$y_0 = cx + \max_{s \in T_p} u^s(b - Ax).$$

Then $y_0 \geqslant cx + u^s(b - Ax)$ for all $s \in T_P$ and we may write P in the form

$$v(P_B) = \min y_0 \qquad\qquad\qquad\qquad\qquad\qquad (P_B)$$

subject to

$$y_0 + u^s Ax - cx \geqslant u^s b \qquad \text{for all } s \in T_P,$$
$$v^r Ax \geqslant v^r b \qquad \text{for all } r \in R_P,$$
$$x \geqslant 0 \text{ and integer.}$$

In the problem P_B, we have only one continuous variable. It follows from the above consideration that the following theorem holds:

THEOREM 6.3. *The problem P_B is equivalent to the problem P.* □

The constraints (6.12) corresponding to extreme rays of $F(DPx)$ may be eliminated if we add in P the extra constraint

$$\sum_{j=1}^{n} x_j \leqslant M, \qquad\qquad\qquad\qquad\qquad\qquad (6.13)$$

with M a sufficiently large positive number. Then in P_B we will have constraints corresponding to the vertices of $F(DPx)$, which we will call the *vertex constraints* or the *primal cuts*.

In general, solving P_B is not easier than solving P as we have to know all vertices of $F(DPx)$. The idea of Benders' decomposition consists in generation of primal cuts as they are needed. Computational experiments indicate that, in general, substantially less than $|T_P|$ primal cuts are needed to solve P.

We will now give a formal description of Benders' decomposition. Without loss of generality, we may assume that an optimal solution $(\bar{x}, \bar{y}) \in F^*(\bar{P})$ is not feasible for P, i.e., that $x \notin Z^n$.

Benders' Method

Step 0 (Initialization): $T_P := \emptyset$; $\underline{z} := -\infty$; $\bar{z} := +\infty$; Select x' according to (6.13);

Step 1 (Linear Programming Phase): Solve DPx'; (Let $u^s \in F^*(DPx')$; $T_P := T_P \cup \{s\}$; If $\bar{z} > v(DPx')$, then $\bar{z} := v(DPx')$;

Step 2 (Integer Programming Phase): Solve P_B; $\underline{z} := v(P_B)$; (Let $x'' \in F^*(P_B)$);

Step 3 (Stop Test): If $\underline{z} < \bar{z}$, then $x' := x''$ and go to Step 1, otherwise stop ($v(P) := \underline{z}$; $x^* := x''$; y^* is an optimal solution to Px^*). □

Benders' method may be considered as a near-optimal method since at Step 1 $(x', y) \in F(P)$, where y is given by the simplex tableau for DPx'. So we may interrupt the computations and consider (x', y) as a near optimal solution of P. Moreover, $\underline{z}$ and $\bar{z}$ are lower and upper bounds on $v(P)$. We also note that at Step 2 we may find only a near-optimal solution $x'' \in F(P_B)$ such that the corresponding y_0 is smaller than $\underline{z}$ computed at the previous iteration.

6.5. THE PRIMAL-DUAL DECOMPOSITION

Integer variables in the problem P studied in the previous section may be considered as complicated ones because, in general, it is much easier to solve a linear programming problem than the corresponding integer programming problem. So in Benders' decomposition we first partition variables into easy ones, i.e., continuous and complicated ones, i.e., integer variables, and next we solve the linear programming problem and the integer programming problem in an alternative way. In this sense, Benders' decomposition may be considered as a primal decomposition.

In the same way, we may consider the constraints of P, namely, we may partition them into "easy" and "hard" ones. Such an approach is called the *dual decomposition*. The partition of constraints is a matter of judgement and depends on available software. But in many practical problem such a partition is unique, for instance, the constraints of type 2 are obviously hard, while $0 \leqslant x \leqslant d$ are easy constraints. Similarly, in the simple plant location problem, the transportation constraints are easy ones. The main rule in such a partition is that the problem obtained after putting the "hard" constraints into the objective function should be much easier to solve than a given problem P.

Combining Benders' decomposition (primal decomposition) with the dual decomposition we get the *primal-dual decomposition*. The aim of this section is to describe its idea. To do this, we formulate P from Section 6.4 in a different way

$$v(P) = \min(cx + dx), \qquad\qquad (P)$$

subject to

$$A_1 x + D_1 y \geqslant b_1,$$
$$A_2 x + D_2 y \geqslant b_2,$$
$$x, y \geqslant 0, \quad x \in Z^n,$$

where b_1 is an m_1-vector and b_2 is an m_2-vector. The other vectors have the same dimensions as in Section 6.4.

We assume that $A_1 x + D_1 y \geqslant b_1$ are "hard" constraints, while $A_2 x + D_2 y \geqslant b_2$ are "easy" ones. Obviously, some matrices in P may vanish and, for instance, the easy constraint may take the form $D_2 y \geqslant b_2$, i.e., the matrix A_2 vanishes.

To avoid some pathological cases, we assume that $F(P) \neq \emptyset$ and bounded. These pathological cases may be considered in a way similar as in Section 6.4. Let u_1 be an m_1-vector and u_2 be an m_2-vector. They form a vector $u = (u_1, u_2)$. Similarly, we denote

$$A = \begin{bmatrix} A_1 \\ A_2 \end{bmatrix}, \quad D = \begin{bmatrix} D_1 \\ D_2 \end{bmatrix}, \quad b = \begin{bmatrix} b_1 \\ b_2 \end{bmatrix}.$$

Consider the Lagrangean relaxation of P in which the hard constraints $A_1 x + D_1 y \geqslant b_1$ are included in the objective function with the coefficient $u_1 \in R^{m_1}$, $u_1 \geqslant 0$:

$$v(Lu_1) = \min(cx + dy + u_1(b_1 - A_1 x - D_1 y)), \qquad\qquad (Lu_1)$$

subject to

$$A_2 x + D_2 y \geqslant b_2,$$
$$x, y \geqslant 0, \quad x \in Z^n.$$

As $F(Lu_1) \supseteq F(P)$ and, for any $u_1 \geqslant 0$, we have $cx + dy + u_1(b_1 - A_1 x - D_1 y)$ for all $(x, y) \in F(P)$, the dual problem to Lu_1 is

$$v(D) = \max_{u_1 \geqslant 0} \min_{x \in F(Lu_1)} (cx + dy + u_1(b - A_1 x - D_1 y)). \tag{D}$$

Let

$$u_0 = \min_{x \in F(Lu_1)} (cx + dy + u_1(b_1 - A_1 x - D_1 y)).$$

Then $cx + dy + u_1(b_1 - A_1 x - D_1 y) \geqslant u_0$ for all $x \in F(Lu_1)$. The problem Lu_1 is a mixed integer programming problem and its objective function attains its minimum at one of the vertices of the convex polytope $\operatorname{conv} F(Lu_1)$, denoted as (x^r, y^r), $r \in T_D$. Then we may write the dual problem as

$$v(D) = \max u_0, \tag{D}$$

subject to

$$u_0 - u_1(b_1 - A_1 x^r - D_1 y^r) \leqslant cx^r + dy^r \quad \text{for any } r \in T_D,$$
$$u_1 \geqslant 0.$$

The constraints in D are called *dual cuts*. Their number equals the number of vertices of $\operatorname{conv} F(Lu_1)$. Note that $F(Lu_1)$ does not depend on the choice of u_1.

Obviously, $v(D) \leqslant v(P)$. Similarly as in Benders' decomposition, we construct dual cuts as they are needed. In the primal-dual decomposition, we solve alternatively the problem P_B formulated in Section 6.4 with (6.14) and problem D. Below we give a formal description of the method.

The Primal-Dual Decomposition Method

Step 0 (Initialization): $k := 0$; $T_P := \emptyset$; $T_D := \emptyset$; $\underline{z} := -\infty$, $\bar{z} := +\infty$; Select $u_1^1 > 0$;

Step 1 (Primal relaxation): $k := k+1$; Solve Lu_1^k (Let $x^k \in F^*(Lu_1^k)$); $T_D := T_D \cup \{k\}$; If $\underline{z} < v(Lu_1^k)$, then $\underline{z} := v(Lu_1^k)$; If $\underline{z} \geqslant \bar{z}$, stop, otherwise go to Step 3b;

Step 2 (Dual relaxation): $k := k+1$; Solve DPx^k (Let $u^k \in F^*(DPx^k)$); $T_P := T_P \cup \{k\}$. If $\bar{z} \geqslant v(DPx^k)$, then $\underline{z} := v(DPx^k)$; If $z \leqslant \bar{z}$, stop, otherwise go to Step 3a;

Step 3 (Stop test):
(3a) Solve D (Let $u_1^{k+1} \in F^*(D)$); Go to Step 1;
(3b) Solve P_B (Let $x^{k+1} \in F^*(P_B)$); Go to Step 2. □

This is the simplest version of the primal-dual decomposition (see references in Section 6.7 for a more developed version of the method).

6.6. THE KNAPSACK PROBLEM AND THE SHORTEST PATH PROBLEM

In this section, we show that an integer programming problem is equivalent to the shortest path problem formulated in a suitable graph. In Section 1.2, we proved that an integer programming problem with bounded variables

$$v(P) = \min\{cx|Ax = b,\ 0 \leqslant x \leqslant d,\ x \in Z^n\} \qquad (P)$$

is equivalent to the (integer) knapsack problem

$$v(K) = \min\left\{\sum_{j=1}^{n} c_j x_j \;\Big|\; \sum_{j=1}^{n} a_j x_j = a_0,\ 0 \leqslant x \leqslant d,\ x \in Z^n\right\}. \qquad (K)$$

Now we show that the above knapsack problem is equivalent to the shortest path problem in the following graph.

Without loss of generality, we may assume that all a_j are different. Consider a directed graph $G = (V, E)$ having $a_0 + 1$ vertices denoted by natural numbers, i.e., $V = \{0, 1, \dots, a_0\}$. Let us define for $j = 1, \dots, n$

$$E_j = \{(i, k)|k - i = a_j, i = 0, 1, \dots, \min\{a_0 - a_j, d_j - 1\},\ i < k \leqslant a_0\}.$$

All arcs from E_j have the same length equal to c_j. Let

$$E = \bigcup_{j=1}^{n} E_j.$$

It is easy to see that G has no cycles and any path from 0 to a_0 corresponds to a feasible solution to K. Then the shortest path from 0 to a_0 corresponds to the optimal solution to K and its length equals $v(K)$. Since G has no cycles and if all $c_j > 0$ for all $j = 1, \dots, n$, the shortest path in G may be found by Dijkstra's algorithm (considered in Section 3.4).

It may be proved that solving the binary knapsack problem by dynamic programming (Section 5.4) is equivalent to finding the shortest path in a graph with $na_0 + 1$ vertices.

6.7. BIBLIOGRAPHIC NOTES

6.1–6.3. The partition of constraints into two types was proposed by Crowder et al. (1983). They also formulated the constraint identification problem and the modification of the lifting procedure. Kianfar (1971, 1976) described the rotation procedure and Walukiewicz (1975) proposed some modifications of it. Kaliszewski and Walukiewicz (1979, 1981) used the rotation procedure in a combination with the method of integer forms. Bradley et al. (1974) described a method for obtaining an equivalent binary constraint based on its minimal covers. Salkin and Breining (1971) consider parallel shifting of a given integer constraint (hyperplane).

6.4–6.5. The primal decomposition was proposed by Benders in 1962 and generalized by Geoffrion in 1971. Modifications of the method are given by Balas and Bergthaller (1983), McDonald and Devine (1977), Sweeney and Murphy (1979),

Faner et al. (1981). Geoffrion and Graves (1974) showed that a combination of Benders' decomposition with the tighter equivalent formulation gives excellent results. The primal-dual decomposition was proposed by Van Roy (1983, 1986).

6.6. Shapiro (1968) proved that the knapsack problem is equivalent to the shortest path problem in a suitably constructed graph. This result can be generalized proving that an integer problem without restriction on sign of basic variables is equivalent to the shortest path in a suitably constructed graph (see Gomory, 1965, Hu, 1970).

6.8. EXERCISES

6.8.1. How, using a surrogate problem, can we check whether a given integer programming problem is infeasible or not?

6.8.2. Consider an open-cast mining problem (Williams, 1974). Assume that an open-cast has a shape of an inverted pyramid of 30 equal cubes. On the surface there are 16 cubes. Successively, on the lower levels there are 9,4 and 1 cubes. Let c_j be the profit obtained from the jth cube, i.e., the difference between the revenue obtained after selling the ore and the cost of extracting it, $j = 1, ..., 30$ (c_j may be negative for cubes near the surface.) Our objective is to maximize the total profit. Formulate this problem and next reformulate it in such a way that the constraint matrix is totally unimodular.

6.8.3. In Exercise 4.6.2 put $x \in \{0, 1\}^4$, find a tighter equivalent formulation and solve it.

6.8.4. Prove Theorem 6.1 (Walukiewicz, 1975).

6.8.5. Show that a hyperplane $a_1 x_1 + ... + a_n x_n = b$, with integer coefficients, passess through an integer point if and only if the greatest common diviser divides b. Use this result in construction of a tighter equivalent formulation (Salkin and Breining, 1971).

6.8.6. Prove that (6.4) is equivalent to $ax \leq a_0, x \in \{0, 1\}^n$, if $\bar{a}_j$ and $\bar{a}_0$ are solutions to the following system of linear inequalities:

$$\sum_{j \in S} \bar{a}_j \geq \bar{a}_0 + 1 \quad \text{for any} \quad S \in \bar{S},$$

$$\bar{a}_0, \bar{a}_j \geq 0, \quad j \in N,$$

where S is a minimal cover for (6.4) and $\bar{S}$ is the family of all minimal covers for (6.4) (Bradley et al., 1974).

6.8.7. Demonstrate that Theorem 6.2 gives a minimal cover S^* such that

$$\sum_{j \in S^*} \bar{x}_j - (|S^*| - 1) = \max_{S \in \bar{S}} \left(\sum_{j \in S} \bar{x}_j - (|S| - 1) \right),$$

where $\bar{S}$ is the family of all minimal covers for (6.6).

6.8.8. For the problem P from Exercise 6.8.3, find a minimal cover inequality cutting off $\bar{x}$ from $F(P)$.

6.8.9. Using Benders' decomposition solve the problem

$$\begin{aligned}
v(P) = \min(2x_1 + 2x_2 + 5y), \\
3x_1 + 3x_2 + y \geqslant 5, \\
-x_1 + x_2 + 4y \geqslant 7, \\
x_1 - x_2 + 2y \geqslant 4, \\
x_1, x_2, y \qquad\qquad \geqslant 0, \quad x \in Z^2.
\end{aligned}$$

Give a graphical interpretation of the method.

6.8.10. Prove that in Benders' decomposition at Step 1 we construct a different vertex of $F(DPx)$ in every iteration. Show that the same is true for x'' constructed at Step 2 of Benders' decomposition.

6.8.11. Prove that the primal-dual decomposition is convergent.

6.8.12. Apply the primal-dual decomposition to the simple plant location problem.

6.8.13. Show that the binary knapsack problem is equivalent to the shortest path problem in a graph with $na_0 + 1$ vertices.

CHAPTER 7

Relaxations of Integer Problems. Duality

In Section 1.6, we demonstrated that the majority of the integer programming problems known so far are difficult to solve; more precisely, they are $\mathcal{NP}$-hard. By including the so-called hard constraints into the objective function, we obtain a Lagrangean relaxation of a given integer programming problem which is far more easy to solve. Although, in general, the solution of a Lagrangean relaxation is not a solution to a given problem, it may serve as a good estimation of it in a branch-and-bound method. Moreover, these estimations are usually better than the bounds obtained by linear programming (see Chapter 4) in the sense that they are tighter and/or they require a lower computational load.

In Section 7.1, we consider Lagrangean relaxations, while in the next section, we discuss some properties of surrogate problems. The applications of these relaxations to the so-called generalized assignment problem are discussed in Section 7.3. In other words, in the first three sections we consider numerical aspects of the integer programming duality.

The rest of this chapter is devoted to theoretical aspects of integer programming duality. We show that it is possible to modify the notion of prices in such a way that the duality theorems known from linear programming (see Section 2.6) hold in the case of integer programming problems.

7.1. LAGRANGEAN RELAXATIONS

Consider a mixed integer programming problem

$$v(P) = \min cx, \qquad (P)$$

subject to

$$Ax \geqslant b,$$
$$Gx \geqslant h,$$
$$x \geqslant 0,$$

$$x_j \text{ integer for all } j \in N_c \subseteq N \ \{1, \ldots, n\},$$

where $x \in R^n$, $b \in R^m$, $h \in R^q$ and all matrices are of proper dimensions.

We assume that the constraints $Ax \geqslant b$ are hard and $Gx \geqslant h$ are easy. For a given $u \in R^m$, $u \geqslant 0$, a *Lagrangean relaxation* of P is an integer programming problem of the form

$$v(Lu) = \min \left(cx + u(b - Ax) \right), \qquad (Lu)$$

with restrictions

$$Gx \geqslant h,$$

$$x \geqslant 0, \quad x_j \text{ integer for all } j \in N_c.$$

The problem Lu is a relaxation of P because $F(Lu) \supseteq F(P)$, and for all $u \geqslant 0$, $x \in F(P)$, we have $cx + u(b - Ax) \leqslant cx$ since $u(b - Ax) \leqslant 0$.

If in P there are equality constraints $Ax = b$, then u is unrestricted in sign. If in analysis of P we wish to omit the ith constraint, then we set $u_i = 0$ and consider the problem Lu. The vector $u \in R^m$ is called a *Lagrange multiplier vector*. Observe also that $F(Lu)$ does not depend on the choice of u.

7.1.1. *Properties of a Lagrangean Relaxation*

The partition of the constraints of P into hard $Ax \geqslant b$ and easy $Gx \geqslant h$ should be such that Lu is much easier to solve than P. For instance, in Section 5.1 we showed that the solution of a Lagrangean relaxation of a given knapsack problem may be computed in $O(n)$, while the knapsack problem is $\mathcal{NP}$-hard.

The usefulness of a Lagrangean relaxation heavily depends on how close $v(Lu)$ is to $v(P)$. Since Lu is a relaxation of P, $v(Lu) \leqslant v(P)$, and therefore one should chose $u \in R^m$, $u \geqslant 0$ in such a way that u is a solution to the following dual problem:

$$v(D) = \max_{u \geqslant 0} v(Lu) = \max_{u \geqslant 0} \ \min_{u \in F(Lu)} (cx + u(b - Ax)). \tag{D}$$

The difference $v(P) - v(D)$ is called the *duality gap*. A simple numerical example shows that in integer programming there does not always exist a $u^* \in F^*(D)$ such that $v(P) = v(D) = v(Lu^*)$, i.e., then we have no duality gap. On the other hand, in linear programming, there exists a $u^* = \bar{u}$ (see Section 2.6) for which we have no duality gap.

To study the duality gap in integer programming, we consider the following linear programming problem:

$$v(P') = \min cx, \tag{P'}$$

subject to

$$x \in \text{conv} \{x \in R^n | Gx \geqslant h, x \geqslant 0, x_j \in Z, j \in N_c\}.$$

Let u' be an optimal solution to P'. To avoid pathological cases, we assume that $F(P')$ is bounded. As usual, $\bar{P}$ denotes the linear programming relaxation of P, i.e., a linear programming problem obtained after dropping the constraint $x_j \in Z, j \in N_c$. Let $\bar{D}$ be the dual problem to $\bar{P}$ and let $(\bar{u}, \bar{v}) \in F^*(D)$, where $\bar{v}$ corresponds to the constraints $Gx \geqslant h$.

THEOREM 7.1.

(a) $F(\bar{P}) \supseteq F(P') \supseteq F(P)$ *and* $F(Lu) \supseteq F(P)$ *for all* $u \geqslant 0$,

$v(\bar{P}) \leqslant v(P') \leqslant v(P)$ *and* $v(Lu) \leqslant v(P)$ *for all* $u \geqslant 0$.

(b) *If* $F(\bar{P}) \neq \emptyset$, *then* $v(\bar{P}) \leqslant v(L\bar{u})$.

(c) *If, for a given* u, *a vector* x *satisfies three conditions*:

$$x \in F^*(Lu),$$ (i)

$$Ax \geqslant b,$$ (ii)

$$u(b - Ax) = 0,$$ (iii)

then $x \in F^*(P)$. If x satisfies (i) and (i), but not (iii), then x is a near-optimal solution to P and the error is $\varepsilon = v(P) - v(Lu) = u(Ax - b)$.

(d) If $F(P') \neq \varnothing$, then

$$v(D) = \max_{u \geqslant 0} v(Lu) = v(Lu^*) = v(P').$$

Proof. We proved part (a) many times in this book and for the sake of completness we will repeat it here.

In proving part (b), we note that the duality theory of linear programming gives (see Section 2.6)

$$v(\overline{P}) = \max_{u \geqslant 0} v(\overline{Lu}) = v(\overline{L\bar{u}}) \leqslant v(L\bar{u}).$$

The last inequality follows from $F(Lu) \subseteq F(\overline{Lu})$ for all $u \geqslant 0$ ($\overline{Lu}$ denotes the linear programming relaxation of Lu). Thus a Lagrangean relaxation may provide a better bound than linear programming and selecting $u = \bar{u}$ we guarantee a bound at least as good as the linear programming one.

As for part (c) we note that (i) and (ii) are implied by $x \in F(P)$ and if (iii) does not hold, cx differs from $v(P)$ by $u(Ax - b)$. Therefore, a Lagrangean relaxation may be a source of near optimal solutions to P.

The first equality in part (d) follows from the definition of problem D. The problem Lu is a linear mixed integer programming problem and such a problem is equivalent to the linear programming problem in which $F(Lu)$ is substituted by $\operatorname{conv} F(Lu)$. Thus P' and D form a pair of dual problems. $\qquad \square$

COROLLARY 7.1. $v(\overline{P}) \leqslant v(L\bar{u}) \leqslant v(Lu^*) = v(D) = v(P') \leqslant v(P), u^* \in F^*(D).$ $\qquad \square$

In other words, a Lagrangean relaxation cannot give better bounds than $v(P')$. Therefore, we study two extreme cases: $v(P') = v(\overline{P})$ and $v(P') = v(P)$.

We say that a Lagrangean relaxation of a given integer problem has the *integrality property* if $v(Lu) = v(\overline{Lu})$ for all $u \geqslant 0$. So, if a Lagrangean relaxation has the integrality property, it is equivalent to a linear programming problem obtained from Lu by dropping integrality requirements $x_j \in Z, j \in N_c$.

THEOREM 7.2. *If $F(\overline{P}) \neq \varnothing$ and Lu has the integrality property, $F(P') \neq \varnothing$ and*

$$v(\overline{P}) = v(L\bar{u}) = v(Lu^*) = v(D) = v(P').$$

Proof. By Corollary 7.1 it is sufficient to prove $v(\overline{P}) = v(P')$.

$$v(\overline{P}) = \max_{u \geqslant 0} v(\overline{Lu}) \qquad \text{by the strong duality theorem,}$$

$$= \max_{u \geqslant 0} v(Lu) \qquad \text{by the integrality property,}$$

$$= \max_{u \geqslant 0} \min \{cx + u(b - Ax) | x \in \operatorname{conv} F(Lu)\} \quad \text{by definition of } Lu,$$

$$= v(P') \qquad\qquad\qquad \text{by duality, Theorem 7.1.}$$

In the same way we show that $F(P) \neq \emptyset$ implies $F(P') \neq \emptyset$. $\qquad\qquad\square$

This theorem does not say that it is impractical to solve a Lagrangean relaxation in the case when it has the integrality property, as in many cases, solving Lu may be much easier than solving $\bar{P}$.

The second extreme case $(v(P') = v(P))$ is much more complicated. The sufficient condition for $v(P') = v(P)$ is $F(P') = \operatorname{conv} F(P)$, but this criterion is impractical as up to now we do not know how to construct $\operatorname{conv} F(P)$ in an efficient way for a given general integer problem.

7.1.2. Calculating Lagrangean Multipliers

The optimal values for Lagrangean multipliers are obtained by solving the problem

$$v(D) = \max_{u \geqslant 0} v(Lu) = \max_{u \geqslant 0} \min_{x \in F(Lu)} \big((cx + u(b - Ax)) \big). \tag{D}$$

Consider $v(Lu)$ as a function of $u \in R^m$, i.e., $f(u) = v(Lu)$. For simplicity, we assume that $N_c = N$ and $F(P) \neq \emptyset$. Then P, as a (pure) integer problem, has a finite number of feasible points, i.e., $F(Lu) = \{x_r | r \in T\}$. So we may write D as

$$v(D') = \max u_0, \tag{D'}$$

subject to

$$u_0 - u(b - Ax_r) \leqslant cx_r \quad \text{for all } r \in T,$$

$$u \geqslant 0.$$

Every constraint in D' is equivalent to the inequality $u_0 \leqslant cx_r + u(b - Ax_r)$. The variable u_0 is unrestricted in sign. In this formulation, the problem D' is dual to the following problem:

$$v(P'') = \min \sum_{r \in T} w_r cx_r \tag{P''}$$

subject to

$$\sum_{r \in T} w_r Ax_r \geqslant b,$$

$$\sum_{r \in T} w_r = 1,$$

$$w_r \geqslant 0 \quad \text{for all } r \in T.$$

Note that P'', with requirements $w_r = 0$ or 1, $r \in T$, is equivalent to P as then x_r satisfies $Gx \geqslant b$ by definition and $x_r \in Z^n$, $x \geqslant 0$ for all $r \in T$. But, in general, the problem P'' is not equivalent to P.

In Fig. 7.1 we present an example of a function $f(u) = v(Lu)$ for $m = 1$ and $|T| = 4$. In Section 7.8, we give references to the following

THEOREM 7.3. $f(u)$ is a continuous, piecewise linear concave function of $u \in R^m$. $f(u)$ is differentiable in any point of R^m but not at points u where Lu has many optimal solutions.   □

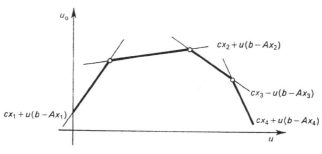

Fig. 7.1.

An optimal solution u^* may be found by subgradient methods described in Section 2.10. Then

$$f(u) \leqslant f(u') + g(u')(u - u'), \tag{7.1}$$

where $g(u')$ is a subgradient of $f(x)$ at u'. It is easy to see that $b - Ax'$ is a subgradient of $f(u)$ at a point u such that $x' \in F^*(Lu)$. From Chapter 2 we know that $u^* \in F^*(D)$ and $w^* \in F^*(P'')$ if and only if u^* and w^* satisfy the complementarity slackness, which is equivalent to the condition that $0 \in R^m$ is a subgradient of $f(u)$ at u^* (Exercise 7.9.4).

So having a starting point u_0, e.g., $u_0 = 0$, we may solve problem D by calculating a sequence of points

$$u_{k+1} = u_k + \gamma_k(b - Ax_k), \tag{7.2}$$

where γ_k is the step length. It may be shown that under the assumption

$$\gamma_k \to 0 \quad \text{and} \quad \sum_{k=1}^{\infty} \gamma_k \to \infty, \tag{7.3}$$

we have $u_k \to u^*$ and $f(u_k) \to v(D)$.

Usually, γ_k is computed as

$$\gamma_k = \frac{\beta_k(\bar{z} - f(u_k))}{\|b - Ax_k\|}, \tag{7.4}$$

where β_k is a parameter $0 \leqslant \beta_k \leqslant 2$ and $\bar{z}$ is an upper bound on $v(D)$, with $x_k \in F^*(Lu_k)$.

Solving problem D by (7.2) requires infinitely many iterations. In practice, the computations are interrupted if in S subsequent iterations no substantial improvement has been made, e.g., if $|f(u_k) - f(u_{k+i})| < \varepsilon$ for a given $\varepsilon > 0$ and $1 \leqslant i \leqslant S$. In particular, if $Ax_k \geqslant b$, then $x_k \in F^*(P)$ and $u_k \in F^*(D)$.

7.1.3. *Using Lagrangean Relaxation in Branch-and-Bound Methods*

We have shown that $v(Lu^*) \geqslant v(\overline{P})$, i.e., a Lagrangean relaxation provides bounds no worse than bounds obtained by linear programming. Computational experiments indicate that usually $v(Lu) > v(\overline{P})$ even for nonoptimal Lagrangean multipliers (as we mentioned above, we compute u^* in a near optimal way). Moreover, even in the case $v(Lu^*) = v(\overline{P})$, solving Lu may be easier than solving $\overline{P}$. As an example, we may consider the knapsack problem and, as we know from Chapter 5, solving a Lagrangean relaxation is easier than solving $\overline{K}$, and in this case the Lagrangean relaxation has the integrality property.

A solution to Lu satisfies the integrality constraints, but it may violate the constraints $Ax \geqslant b$. In some practical cases, such a solution may be of some practical value, even greater than that of $\overline{x} \in F^*(\overline{P})$. This is the case when the constraints $Ax \geqslant b$ may be considered as soft ones, i.e., when some violation of them is acceptable, while the requirements $x_j \in Z, j \in N_c$, are important.

Lagrangean relaxations are used in branch-and-bound methods in exactly the same way as we use linear programming in Chapter 4. In particular, $v(Lu)$ may be used to estimate up and down penalties, to select a subproblem from the list of subproblems and to select a branching variable. Combining Lagrangean relaxation with linear programming usually gives good results. For instance, if $\overline{x} \in F^*(\overline{P})$ and $x_j \in Z$ for some $j \in N_c$, instead of considering two linear programming problems defined by (4.7) and (4.8), respectively, we solve two Lagrangean relaxations $v(L\overline{u}|x_j \leqslant \lfloor \overline{x}_j \rfloor)$ and $v(L\overline{u}|x_j \geqslant \lfloor \overline{x}_j \rfloor + 1)$, where $\overline{u} \in F^*(\overline{D})$ and $\overline{D}$ is the dual problem to $\overline{P}$.

7.2. SURROGATE PROBLEMS

Let $X = \{x \in R^n | Gx \geqslant h, x \geqslant 0, x_j \in Z, j \in N_c\}$. Then P may be written as

$$v(P) = \min\{cx | Ax \geqslant b, x \in X\}, \qquad (P)$$

where $b \in R^m$. For a given $w \in R^m, w \geqslant 0$, a *surrogate problem* is defined as follows:

$$v(Sw) = \min\{cx | wAx \geqslant wb, x \in X\}, \qquad (Sw)$$

i.e., in a surrogate problem, we substitute $m > 1$ constraints by one constraint $wAx \geqslant wb$. In the case when G vanishes and $N_c = N$, i.e., when $X = \{x \in R^n | x \geqslant 0, x \in Z^n\}$, the surrogate problem is the knapsack problem.

The problem Sw is a relaxation of P as $F(Sw) \supseteq F(P)$ and the objective function is the same in both problems. If P is an $\mathcal{NP}$-hard problem, then Sw is also $\mathcal{NP}$-hard, and one may ask what is the advantage of solving Sw instead of P. These advantages are easily seen when Sw is the knapsack problem. As we know from Chapter 5, analysis and solution methods of the knapsack problem are far more developed and efficient than in the case of a general integer programming problem. For instance, $F(Sw) = \emptyset$ implies $F(P) = \emptyset$. This criterion is useful in branch-and-bound methods.

Since Sw is a relaxation, it is reasonable to formulate a *surrogate dual problem*

$$v(D_s) = \max_{w \geqslant 0} v(Sw) = \max_{w \geqslant 0} \min_x \{cx | wAx \geqslant wb, x \in X\}. \qquad (D_s)$$

Similarly as in the case of the Lagrangean relaxation, we define the *duality gap* (*for a surrogate problem*) as $v(P) - v(D_s)$. A simple numerical example shows that, in general, we have $v(P) - v(D_s) > 0$.

In Section 1.2, we showed that $m > 1$ equality constraints may be aggregated into one equality constraint in such a way that the feasible solution set is preserved. The aggregated problem may be considered as the surrogate problem and in this case we have $v(Sw) = v(P)$, i.e., we have no duality gap. The above remarks indicate that, even in the case of inequality constraints, one should expect better bounds than the ones obtained by Lagrangean relaxations.

THEOREM 7.4. *If Lu and Su are, respectively, a Lagrangean relaxation and a surrogate problem for a given problem P and for given $u \in R^m$, $u \geqslant 0$, then $v(Su) \geqslant v(Lu)$.*

Proof. By definition we have

$$v(Lu) = \min \{cx + u(b - Ax) | x \in X\}, \qquad (Lu)$$
$$v(Su) = \min \{cx | uAx \geqslant ub, x \in X\}, \qquad (Su)$$

so $F(Su) \subseteq F(Lu)$, and for any $x \in F(Su)$

$$cx \geqslant cx + u(b - Ax) \qquad \text{since then } ub \leqslant uAx. \qquad \square$$

Thus, a surrogate problem gives, in general, better bounds than a Lagrangean relaxation, both of which were constructed for the same multiplier vector $u \in R^m$, $u \geqslant 0$. We may further improve this bound if we solve problem D_s. As a consequence of Theorem 7.2 and Corollary 7.1, we get

COROLLARY 7.2. $v(\overline{P}) \leqslant v(D) \leqslant v(D_s) \leqslant v(P)$. $\qquad \square$

Similarly as in the case of Lagrangean relaxation, we say that Sw has the integrality property if $v(Sw) = v(\overline{Sw})$ for all $w \in R^m$, $w \geqslant 0$, where, as usual, $\overline{Sw}$ denotes the linear programming relaxation of Sw. The problem $\overline{Sw}$ is called the *linear surrogate problem*. We observe that, in general, a surrogate problem does not have the integrality property as often as a Lagrangean relaxation. For instance, if a surrogate problem is the knapsack problem K, only for very particular data do we have $v(K) = v(\overline{K})$.

To simplify notations, we consider for a moment the problem

$$v(P) = \min \{cx | Ax \geqslant b, 0 \leqslant x \leqslant d, x \in Z^n\}. \qquad (P)$$

Then, for given $w \in R^m$, $w \geqslant 0$, we get

$$v(Sw) = \min \{cx | wAx \geqslant wb, 0 \leqslant x \leqslant d, x \in Z^n\}, \qquad (Sw)$$

and the linear surrogate problem takes the form

$$v(\overline{Sw}) = \min \{cx | wAx \geqslant wb, 0 \leqslant x \leqslant d\}, \qquad (\overline{Sw})$$

so it is the linear knapsack problem. Obviously, we have $v(\overline{Sw}) \leqslant v(Sw) \leqslant v(P)$, and it is reasonable to formulate the problem

$$v(\overline{D}_s) = \max_{w \geqslant 0} v(\overline{Sw}). \tag{$\overline{D}_s$}$$

This problem is far more easy than D_s and its solution is given by

THEOREM 7.5. *If $\overline{D}$ is the dual problem to $\overline{P}$ and if $\overline{u} \in F^*(D)$, then $\overline{w} = \overline{u}$ is an optimal solution to $\overline{D}_s$.*   □

Surrogate problems are used in branch-and-bound methods in exactly the same way as Lagrangean relaxations. Usually, the problem D_s is solved in a near-optimal way by a subgradient methods. We may exclude $0 \in R^m$ from the optimal solution set $F^*(D_s)$ as $v(Sw|w = 0) \leqslant v(Sw)$ for all $w \geqslant 0$. Additionally, $v(Sw) = v(Skw)$ for any $k > 0$, and therefore we may normalize w, e.g., in the following way:

$$\sum_{i=1}^{m} w_i = 1.$$

7.3. THE GENERALIZED ASSIGNMENT PROBLEM

Consider a problem of assigning "jobs" $J_1, \ldots, J_n$ to "machines" $M_1, \ldots, M_m$. Let c_{ij} be a cost of performing job J_j on M_i and let b_i be a resource of machine M_i. By a_{ij} we denote an execution time of job J_j on machine M_i. Let $x_{ij} = 1(0)$ if job J_j is (is not) assigned to machine M_i. The so-called *generalized assignment problem* may be formulated as

$$(P) \qquad v(P) = \min \sum_{i=1}^{m} \sum_{j=1}^{n} c_{ij} x_{ij}, \tag{7.5}$$

subject to

$$\sum_{j=1}^{n} a_{ij} x_{ij} \leqslant b_i, \quad i = 1, \ldots, m, \tag{7.6}$$

$$\sum_{i=1}^{m} x_{ij} = 1, \qquad j = 1, \ldots, n, \tag{7.7}$$

$$x_{ij} = 0 \text{ or } 1, \quad i = 1, \ldots, m, \ j = 1, \ldots, n. \tag{7.8}$$

Constraints (7.6) corespond to limits on working time for each machine M_i, while (7.7) means that each job may be performed only on one machine. From the formulation of the problems, we know that all data are nonnegative integers.

There are two natural Lagrangean relaxations for the generalized assignment problem. The first is obtained by dualizing constraints (7.6) for a given $u \in R^m$, $u \geqslant 0$:

$$v(Lu) = \min \left(\sum_{i=1}^{m} \sum_{j=1}^{n} c_{ij} x_{ij} + \sum_{i=1}^{m} u_i \left(\sum_{j=1}^{n} a_{ij} x_{ij} - b_i \right) \right) \tag{Lu}$$

under constraints (7.7) and (7.8). It can be written as

$$v(Lu) = -\sum_{i=1}^{m} u_i b_i + \min\left(\sum_{j=1}^{n}\left(\sum_{i=1}^{m} c_{ij} + u_i a_{ij}\right)x_{ij}\right).$$

The relaxation Lu may be easily solved by determining a minimal element in each column of matrix $c_{ij} + u_i a_{ij}$, Then the corresponding $x_{ij}^* = 1$, while all remaining x_{ij}^* in this column are set to zero. Computing a minimal element in a column requires $m-1$ comparisons. Thus Lu may be solved in time proportional to mn using $n(m-1)$ comparisons.

The second relaxation of P is obtained by dualizing constraints (7.7) for a given $q \in R^n$:

$$v(Lq) = \min\left(\sum_{i=1}^{m}\sum_{j=1}^{n} c_{ij}x_{ij} + \sum_{j=1}^{n} q_j\left(\sum_{i=1}^{m} x_{ij} - 1\right)\right), \qquad (Lq)$$

subject to (7.6) and (7.8). The objective function of Lq may be written as

$$v(Lq) = -\sum_{j=1}^{n} q_j + \min\left(\sum_{i=1}^{m}\sum_{j=1}^{n} (c_{ij} + q_j)x_{ij}\right).$$

So Lq decomposes into m binary knapsack problems. As we know from Chapter 5, the knapsack problem may be efficiently solved although it is $\mathcal{NP}$-hard. For instance, it can be solved by dynamic programming by $O(nb_i)$ additions and comparisons. Therefore, Lq may be solved in

$$O\left(n \sum_{i=1}^{m} b_i\right)$$

additions and comparisons.

Usually, $q = 0$, with $q \in R^n$, is a natural starting point to a subgradient method for computing $q^* \in F^*(D)$. But this starting point may be selected in a more efficient way. If $c_{ij} > 0$ for all i, j, then $x_{ij} = 0$ for all i, j is an optimal solution to Lq, providing $|q_j| \leqslant c_{ij}$. Therefore, selecting

$$q_j = \min_i c_{ij}, \quad j = 1, \ldots, n,$$

is a better starting point than $q = 0$.

Usually, there is a trade-off between the quality of bounds and the computational load necessary compute them. For instance, a relaxation which is harder to solve might provide better bounds. For the generalized assignment problem, the relaxation Lu has the integrality property, while Lq has not. Therefore, by Theorem 7.2,

$$v(P) \geqslant v(Lq^*) \geqslant v(Lu^*) = v(L\bar{u}) = v(\bar{P}),$$

where $\bar{u} \in F^*(\bar{D})$ and $\bar{D}$ is the dual problem to $\bar{P}$. Comparing computational complexities of Lu and Lq we see that Lq is far more difficult to solve than Lu. Computational experiments indicate that this is true (see Section 5.8).

Consider now a surrogate problem for a given $w \geqslant 0$

$$v(Sw) = \min \sum_{i=1}^{m}\sum_{j=1}^{n} c_{ij}x_{ij},$$

subject to

$$\sum_{i=1}^{m} w_i \sum_{j=1}^{n} a_{ij} x_{ij} \leqslant \sum_{i=1}^{m} w_i b_i,$$

$$\sum_{i=1}^{m} x_{ij} = 1, \quad j = 1, \ldots, n,$$

$$x_{ij} = 0 \text{ or } 1, \quad i = 1, \ldots, m, j = 1, \ldots, n.$$

The problem Sw is the multiple-choice knapsack problem (Section 5.6). By Theorem 7.4, we have $v(S\bar{u}) \geqslant v(L\bar{u}) = v(\bar{P})$. We may reduce the number of variables in both Lq and Sw using the methods described in Chapter 5.

7.4. PRICE FUNCTIONS

So far we have been solving problems D or D_s in order to obtain bounds on $v(P)$, or, in other words, we have considered computational aspects of integer programming duality. We have shown that, in general, the dual problems D and D_s do not give a zero duality gap, therefore the strong duality theorem does not hold (see Section 2.6) and in consequence the corollaries from this theorem are not valid in the case of the integer programming problem. In other words, we have neither the necessary nor sufficient conditions for the optimality for integer problems or complementarity slackness. Recently, much work has been done to construct an anolog of the linear programming duality theory for the case of integer problems. In the rest of this chapter, we will consider some aspects of such a theory connected with a generalization of the notion of "dual variable" introduced in Section 2.6.

From the methodological point of view, it is better to consider first the duality theory for general mathematical programming and next the integer programming duality as a particular case of such a theory.

Let

$$v(P) = \sup_{x} \{f(x)|g(x) \leqslant b, x \in X\} \qquad (P)$$

be a general integer programming problem with $b \in R^m$. If f and g are linear functions and $X = \{x \in R^n | x \geqslant 0, x \in Z^n\}$, then P is an integer programming problem. In this formulation, we have the operator "sup" as we do not assume that $f(x)$ attains its maximum on $F(P)$, i.e., we do not assume $F^*(P) \neq \varnothing$.

Let H_+^m be a set of all nondecreasing functions $h: R^m \to R \cup \{+\infty, -\infty\}$, i.e., functions such that if $d_1 \leqslant d_2$, then $h(d_1) \leqslant h(d_2)$. Any element of H_+^m is called a *price function*.

For any $H \subseteq H_+^m$, we define the dual problem to P as

$$v(D) = \inf_{h} h(b), \qquad (D)$$

subject to

$$h(g(x)) \geqslant f(x) \quad \text{for all } x \in X,$$

$$h \in H.$$

Example 7.1. Let P be a linear programming problem

$$v(P) = \max\{cx|Ax \leqslant b, x \geqslant 0\} \qquad\qquad (P)$$

and let H be now a set of all nondecreasing linear price functions. Then each $h \in H$ may be written as

$$h(d) = \sum_{i=1}^{m} u_i d_i,$$

where $u \in R^m$ and $u \geqslant 0$. Then the problem D for this case takes the form

$$v(D) = \inf ub,$$
$$uAx \geqslant cx \qquad \text{for all } x \geqslant 0,$$
$$u \geqslant 0.$$

Observe that, for $x \geqslant 0$, the set $\{u \geqslant 0|uAx \geqslant cx\} = \{u \geqslant 0|uA \geqslant c\}$ and thus

$$v(D) = \inf\{ub|uA \geqslant c, u \geqslant 0\}. \qquad\qquad (D)$$

So it is exactly the same as the dual problem considered in Section 2.6.   □

By our assumption in Section 2.6, $v(P) = -\infty$ if $F(P) = \emptyset$, and $v(D) = +\infty$ if $F(D) = \emptyset$. Now we show that, for such a pair, P and D, theorems similar to ones proved in Section 2.6 hold.

THEOREM 7.6. (*Weak Duality*). $f(x) \leqslant h(b)$ *for all* $x \in F(P)$ *and all* $h \in F(D)$.

Proof. If $F(P) = \emptyset$ and/or $F(D) = \emptyset$, then by our assumption the theorem holds. If $F(P) \neq \emptyset$ and $F(D) \neq \emptyset$, then $g(x) \leqslant b$ and $h(g(x)) \geqslant f(x)$ for all $x \in X$, providing $x \in F(P)$ and $h \in F(D)$. Since h is nondecreasing, we have

$$f(x) \leqslant h(g(x)) \leqslant h(b).$$

So $f(x) \leqslant f(b)$ for all $x \in F(P)$ and all $h \in F(D)$.   □

7.4.1. *Perturbation Functions*

In sensitivity analysis, it is useful to know how $v(P)$ changes when the right-hand-side b in the constraints $g(x) \leqslant b$ is perturbed (changed).

DEFINITION 7.1. A function $v: R^m \to R \cup \{+\infty, -\infty\}$ is called a *perturbation function* for the problem P if

$$v(y) = \sup\{cx|Ax \leqslant y, x \in X\}. \qquad\qquad □$$

It is easy to prove (Exercise 7.9.17) the following

THEOREM 7.7. *The perturbation function for a linear programming problem*

$$v(y) = \sup\{cx|Ax \leqslant y, x \geqslant 0\}$$

is a continuous piecewise linear convex function.   □

A simple numerical example (Exercise 7.9.18) shows that for an integer programming problem the perturbation function is neither continuous nor convex.

THEOREM 7.8. (*Strong Duality*). *If* $v \in H$, *then* $v(P) = v(D)$.

Proof. By Definition 7.1, v is nondecreasing and $v(g(z)) = \sup\{f(x)|g(x) \leqslant g(z),$ $x \in X\}$ for all $z \in R^n$, where $g: R^n \to R^m$. If $z \in X$, then $z \in \{x|g(x) \leqslant g(z), x \in X\}$, i.e., $z \in F(P)$. Then $v(g(z)) \geqslant f(z)$ for all $z \in X$, and therefore $v \in F(D)$. So $v(b) \geqslant v(D)$. Since $v(P) = v(b)$ and by Theorem 7.6, we have $v(P) \leqslant v(D)$, and $v(P) = v(D)$. $\quad\square$

As a corollary from Theorem 7.8 we get that, by selecting $H \subseteq H_+^m$ in such a way that the perturbation function belongs to H, we guarantee a zero duality gap.

7.4.2. Optimality Conditions and Economic Interpretations

Now we may formulate the necessary and sufficient condition for optimality in a general mathematical programming problem P.

THEOREM 7.9. (*Optimality Conditions*). *A vector* x^* *is an optimal solution to* P *if and only if there exists a pair* (x^*, h^*) *such that*
   (i) $g(x^*) \leqslant b$,
   (ii) $x^* \in X$,
   (iii) $h^*(g(x)) \geqslant f(x)$ *for all* $x \in X$,
   (iv) $h^* \in H_+^m$,
   (v) $h^*(g(x^*)) = h^*(b)$,
   (vi) $f(x^*) = h^*(g(x^*))$.

Proof. If (i)–(vi) hold, then, by Theorem 7.6, $x^* \in F^*(P)$. If $x^* \in F^*(P)$, a function h^* is the perturbation function, and, by Theorem 7.8, conditions (i)–(vi) are satisfied.$\square$

Theorem 7.9 has a limited practical value as, in general, problem D has infinitely many constraints and in the case of integer programming the perturbation function is neither continuous nor convex. But Theorem 7.9 allows economic interpretation for a pair of problems P and D (compare Section 2.6) to be given.

Consider a problem of maximizing production output under limited resources b. Then X is a set of possible production levels and $g(x)$ is a consumption of resources when the production level equals x. The optimal price function h^* has the following properties:
   (a) Price functions are nondecreasing as $h \in H \subseteq H_+^m$.
   (b) No production level makes a positive profit as, by (iii), $f(x) - h^*(g(x)) \leqslant 0$.
   (c) An optimal production level makes a zero profit by (vi).
   (d) The price of the actual consumption vector is not changed by the addition of the unused resources as, by (v),

$$h^*(g(x^*)) - h^*(b) = 0.$$

7.5. DUALITY IN INTEGER PROGRAMMING

Now we consider an application to integer programming of the duality theory developed in the previous section. Let

$$v(P') = \sup\{cx|Ax \leqslant b, x \geqslant 0, x \in Z^n\} \qquad (P')$$

be an integer programming problem with integer data and $b \in R^m$.

DEFINITION 7.2. A function $h \in H_+^m$ is *superadditive* if

$$h(b_1 + b_2) \geqslant h(b_1) + h(b_2) \quad \text{for all } b_1, b_2 \in R^m.$$

If h is superadditive, then $h' = -h$ is *subadditive*.   □

LEMMA 7.1. *The perturbation function v for problem P' is superadditive and $v(0) \geqslant 0$.*

Proof. By Definition 7.1, the perturbation function for P' takes the form

$$v(y) = \sup\{cx|Ax \leqslant y, x \geqslant 0, x \in Z^n\}.$$

Obviously, $v(0) \geqslant 0$. Let x_1 be an optimal solution to P' with the right-hand side b_1. Similarly, $x_2 \in F^*(P|b = b_2)$. Then $x_1 + x_2$ is a feasible solution to P', with the right-hand side $b = b_1 + b_2$. Therefore, $v(b_1 + b_2) \geqslant v(b_1) + v(b_2)$. If $F(P'|b = b_1) = \varnothing$ and/or $F(P'|b = b_2) = \varnothing$, then the theorem also holds.   □

Let now $H \subseteq H_+^m$ be a subset of superadditive and nondecreasing functions with $h(0) \geqslant 0$ for all $h \in H$.

Consider the problem

$$v(D') = \inf h(b), \qquad (D')$$

subject to

$$h(a_j) \geqslant c_j, \quad j = 1, \ldots, n,$$
$$h \in H,$$

where a_j is the jth column of A.

THEOREM 7.10. *The problem D' is dual to P' and $v(P') = v(D')$.*

Proof. We show that D' is a particular case of D defined in Section 7.4. If H is a subset of superadditive functions and $X = \{x|x \geqslant 0, x \in Z^n\}$, then

$$F(D) = \{h \in H|h(Ax) \geqslant cx \quad \text{for all } x \in X\}$$

$$= \left\{h \in H \,\Big|\, h\left(\sum_{j=1}^n a_j x_j\right) \geqslant \sum_{j=1}^n c_j x_j \quad \text{for all } x \in X\right\}$$

$$= \{h \in H|h(a_j)x_j \geqslant c_j x_j, j = 1, \ldots, n\} \quad \text{as } x \geqslant 0, x \in Z^n, h \in H,$$
$$= \{h \in H|h(a_j) \geqslant c_j, j = 1, \ldots, n\} \quad \text{as } x \geqslant 0$$
$$= F(D').$$

Since the perturbation function v of P' belongs to $F(D')$, $v(D') = v(P)$.   □

Observe that in integer programming, similarly as in linear programming, the dual problem does not contain the primal variables. Theorem 7.10 demonstrates one more time how important the notion of the perturbation function is.

It is not difficult to see that in solving the knapsack problem by dynamic programming (Section 5.4) we in fact construct the perturbation function. Moreover, it can be shown that any branch-and-bound method is equivalent to construction (at least partial construction) of the perturbation function.

7.6. REMARKS ABOUT SENSITIVITY ANALYSIS

In typical situations, the data for a mathematical programming problem are collected in the past, while the decisions taken on the basis of an optimal solution are applied in the future. As such data may change in time, e.g., change of costs or capital, one may ask whether such an approach is valid.

In answer to the question, we distinguish two approaches. The first of them consists of considering the data as stochastic variables changing in time. Then we have to solve a stochastic programming problem which, among others, requires determination of distribution functions for the stochastic variables, what in general is not an easy task. Far more popular is the second approach based on the assumption that small changes in data produce small changes in $v(P)$ and $x^* \in F^*(P)$. In other words, for a mathematical model (problem of mathematical programming) to have some practical value, we require it to be stable. Such questions are studied in a chapter of mathematical programming called *sensitivity analysis*.

Sensitivity analysis is well developed in linear programming. Due to the duality theory, one may estimate the influence of a given change on $v(P)$ and $x \in F(P)$ or estimate bounds for data which do not change a given optimal solution x^* to P. In discrete programming, the situation is much more complicated. It is easy to construct an integer programming problem in which relatively small perturbations of data may produce as large as possible changes in $v(P)$ and $x^* \in F^*(P)$. In this section, we show how the duality theory developed in previous sections may help in sensitivity analysis of a given integer programming problem.

Let P be a given integer programming problem and let P' be a problem obtained from P by a perturbation of data. We assume that $x^* \in F^*(P)$, $x' \in F^*(P')$, $h^* \in F^*(D)$ and $h' \in F^*(D')$, where $D(D')$ is the dual problem to $P(P')$. We will consider four cases.

1. *Changing the right-hand-side b into b'*. Then $h^* \in F(D')$ and, in consequence, $v(P') \leqslant h^*(b')$. Let

$$Y^* = \left\{ x \,\middle|\, h^*\left(\sum_{j=1}^{n} a_j x_j\right) = \sum_{j=1}^{n} c_j x_j \right\}.$$

If h^* is still an optimal solution to D', then $x' \in Y^*$.

2. *Changing the coefficients of the objective function c into c'*. Obviously, $x^* \in F(P')$,

and therefore, $v(P') \geq c'x^*$. If $c'_j \leq h^*(a_j)$ for all $j = 1, \ldots, n$, then $h^* \in F(D')$, and in consequence $v(P') \leq h^*(b)$. If $c'_j \leq h^*(a_j)$ when $x^*_j = 0$ and $c'_j = c_j$ when $x^*_j \geq 0$, then $x^* \in F^*(P')$.

3. *Introducing a new variable x_{n+1}*, i.e., we add to the data of P a number c_{n+1} and a vector $a_{n+1} \in Z^m$. Then $(x^*, 0) \in F(P')$, and thus $v(P') \geq v(P)$. The vector $(x^*, 0)$ is an optimal solution to P' if $h^*(a_{n+1}) \geq c_{n+1}$.

4. *Adding a new constraint $\alpha x \leq \beta$*. If $x^* \in F(P')$, then $x^* \in F^*(P')$. The price function h defined in the following way:

$$h(d, d_{m+1}) = h^*(d) + 0 \cdot d_{m+1},$$

is dual feasible, and therefore $v(P') \leq h(b, \beta)$.

7.7. DUALITY AND CONSTRAINT AGGREGATION

Consider a bounded consistent integer programming problem with equality constraints

$$v(P) = \max\{cx | Ax = b, 0 \leq x \leq d, x \in Z^n\}, \tag{P}$$

where $b \in R^m$, and, without loss of generality, we may assume $b \geq 0$. In this section, we will study relations between the aggregation of constraints (see Section 1.2) and duality.

For a given $u \in Z^m$, an *integer surrogate problem* for P is called an integer problem

$$v(Su) = \max\{cx | uAx = ub, 0 \leq x \leq d, x \in Z^n\}. \tag{Su}$$

Problem Su is a relaxation of P as $F(Su) \supseteq F(P)$, and thus $v(Su) \geq v(P)$ for all $u \in Z^m$. For reasons discussed many times in this book, we define the dual problem

$$v(D_s) = \min_{u \in Z^m} v(Su) = \min_{u \in Z^m} \max_{x \in F(Su)} cx. \tag{D_s}$$

Since in P we have equality constraints, a vector of aggregation coefficients $\hat{u} = (\hat{u}_1, \ldots, \hat{u}_m)$ is, obviously, an optimal solution to D_s and the duality gap is zero in this case, i.e., $v(P) = v(D_s)$. Let $\hat{U}$ be a set of all aggregation coefficients for the systems in integers

$$Ax = b, \quad 0 \leq x \leq d. \tag{7.9}$$

Obviously, $k\hat{u} \in \hat{U}$ for any integer k, and moreover $\hat{U} \subseteq F^*(D_s)$. A simple numerical example (Exercise 7.9.18) shows that $\hat{U} \neq F^*(D)$. As the optimal solution set to D_s is unbounded, we may always limit $F(D_s)$, e.g., in the following way:

$$F(D_s) = \{u \in Z^m | 0 \leq u \leq |\hat{u}|\},$$

where $\hat{u}$ is a given aggregation coefficient for (7.9) and $u \leq |\hat{u}|$ means $u_1 \leq |\hat{u}_1| \ldots$ $\ldots, u_m \leq |\hat{u}_m|$.

Let x' be an optimal solution to Su'. There are two possibilities: (1) $x' \notin F(P)$ and (2) $x' \in F(P)$.

(1) If $x' \notin F(P)$, then $Ax' \neq b$ although $u'Ax' = u'b$. Therefore, in the next iteration, we restrict $F(Su')$ by adding one more constraint

$$uAx' \neq ub$$

which due to the integrality of data is equivalent to

$$\text{either} \quad u(b-Ax') \geqslant 1 \quad \text{or} \quad u(b-Ax') \leqslant -1.$$

So we have obtained two subproblems D_s^+ and D_s^- which may be solved in the same way as D_s.

(2) If $x' \in F(P)$, then $x' \in F^*(P)$ and the corresponding vector u' is called an optimal multiplier vector, i.e., $u' \in F^*(D_s)$.

Below we give a description of a branch-and-bound method for solving P and D_s at the same time.

Algorithm A5

Step 0 (Initialization): $k := 0$; $u_k := 0$;

Step 1 (Solution of Su_k): Solve $v(Su_k) = \max\{cx | u_k Ax = u_k b, 0 \leqslant x \leqslant d, x \in Z^n\}$; (Let $x_k \in F^*(Su_k)$). If $Ax_k = b$, stop ($x_k \in F^*(P)$, $u_k \in F^*(D_s)$), otherwise $k := k+1$;

Step 2 (Solution of D_k): Solve $v(D_k) = \min\{cx | uAx_r \neq ub, r = 1, \ldots, k-1, 0 \leqslant u \leqslant |\hat{u}|, u \in Z^m\}$; (Let $u_k \in F^*(D_k)$); go to Step 1.   □

Algorithm A5 is finite as at most

$$\prod_{i=1}^{m} (u+1)$$

iterations are needed to find $x^* \in F^*(P)$ and $u^* \in F^*(D_s)$.

It is possible to give an economic interpretation of a multiplier vector u^* computed by Algorithm A5. Consider again P as a problem of profit maximization. Then $u \in Z^m$ may be taken as prices, and the constraint $uAx = ub$ as a financial restriction on the activity of a firm. The vector $u^* \in F^*(D_s)$ assures that if the financial constraint is satisfied, the resource constraint $Ax \leqslant b$ is satisfied too, and a firm gains maximal profit. Moreover, since $u^* \in Z^m$, it has the same nature as all data in P.

7.8. BIBLIOGRAPHIC NOTES

7.1. Geoffrion (1974) is commonly considered as a basic reference paper on Lagrangean relaxations in integer programming. Different methods for computing Lagrangean multipliers are described by Fisher (1981), Geoffrion (1974), Marsten (1975) and Held et al. (1974). In 1979, Shapiro wrote the first book on Lagrangean relaxations (see also a review paper by Fisher, 1981).

7.2. Glover introduced the notion of surrogate constraint in 1968 (see also his review paper from 1975). Geoffrion proved Theorem 7.5 in 1969. Dyer (1980) described two methods for computing surrogate multipliers and studied the function $f(w) = v(Sw)$ (see also Greenberg and Pierskalla, 1970). Karwan and Rardin (1979) considered relations between Lagrangean and surrogate relaxations.

7.3. Chalmet and Gelders (1977) and Fisher (1981) described the use of Lagrangean relaxations for solving the generalized assignment problem.

7.4 and 7.5. The basic references are papers by Tind and Wolsey (1981) and Wolsey (1981). Properties of the perturbation function are studied in Blair and Jeroslaw (1977) and (1979).

7.6. Sensitivity analysis is considered by Geoffrion and Nauss (1977), Shapiro (1977) and Libura (1977). The last paper contains the sensitivity analysis for the knapsack problem.

7.7. A similar approach consisting of restriction of a feasible solution set is considered by Bell and Shapiro (1977) and Walukiewicz (1981).

7.9. EXERCISES

7.9.1. Construct an integer problem for which $v(D)-v(P) > 0$.

7.9.2. Demonstrate that if in P, $Gx \leqslant h$ reads $0 \leqslant x \leqslant d$, the Lagrangean relaxation has the integrality property.

7.9.3. Prove Theorem 7.3.

7.9.4. Show that complementarity slackness implies $0 \in \partial f(u^*)$.

7.9.5. Prove that if (7.3) holds, the sequences u_k converges to u^* when $k \to \infty$.

7.9.6. Modify methods for computing penalties to the case of Lagrangean relaxations.

7.9.7. Prove Theorem 7.5.

7.9.8. Modify the subgradient method for the case of computing surrogate multipliers (Dyer, 1980).

7.9.9. Incorporate Lagrangean relaxations and surrogate relaxations into a branch-and-bound method.

7.9.10. For given

$$C = \begin{bmatrix} 72 & 38 & 51 & 43 \\ 96 & 21 & 49 & 34 \\ 13 & 52 & 43 & 49 \\ 60 & 77 & 13 & 20 \end{bmatrix}, \quad A = \begin{bmatrix} 11 & 8 & 8 & 9 \\ 14 & 8 & 13 & 8 \\ 3 & 5 & 11 & 8 \\ 7 & 5 & 7 & 5 \end{bmatrix}, \quad b = \begin{bmatrix} 13 \\ 15 \\ 19 \\ 10 \end{bmatrix},$$

formulate the generalized assignment problem and solve it by: (a) Lagrangean relaxation Lu, (b) Lagrangean relaxation Lq, and (c) surrogate relaxation Sw.

7.9.11. Prove Theorem 7.7.

7.9.12. For a given knapsack problem

$$v(K) = \max(3x_1 + 4x_2 + 5x_3 + 3x_4 + x_5 + x_6)$$
$$x_1 + 2x_2 + 5x_3 + 6x_4 + 4x_5 + 10x_6 \leqslant y,$$

study the perturbation functions for K and $\overline{K}$, and compute the values of these functions for $y = 5$.

7.9.13. Using Algorithm A5 find a vector of optimal multipliers for

$$v(P) = \max(15x_1 + 10x_2 + 8x_3)$$
$$10x_1 + 6x_2 + 10x_3 + x_4 \qquad = 41$$
$$4x_1 + 4x_2 + x_3 \qquad + x_5 = 10,$$

$$x \geqslant 0, \quad x \in Z^5.$$

CHAPTER 8

Some Particular Integer Programming Problems

The problems considered in this chapter are particular because they have binary constraint matrices and their right-hand-sides are vectors of ones. We start from the set packing, set partitioning and set covering problems, which have many applications and may also be defined on appropriately constructed graphs. We also investigate relations between these problems and the knapsack problem. In the second part of this chapter, we consider the travelling salesman problem and some of its generalizations.

All such problems can be solved by the general methods described in Chapter 3 and 4. Moreover, the efficiency of the methods may be increased by the constraint generation techniques (Section 6.3) and/or by the reduction of the number of variables in these problems. This is the main objective of this chapter, while the modifications of branch-and-bound methods are mainly considered in the exercise section.

8.1. Packing, Partitioning and Covering Problems

We begin with a general formulation pattern and next show that many practical problems may be formulated according to this pattern.

For $M = \{1, ..., m\}$, let N be a family of "acceptable" subsets of M. We associate a profit (cost) c_j with a member N_j of N and without loss of generality we may assume that $N = \{1, ..., n\}$. The problems considered in this section involve selection of such accepted subsets of M for which the total profit is maximal (the total cost is minimal) and additionally one of the three following conditions is satisfied. Every $i \in M$ is contained:

(a) in at most one of the selected members of N,

(b) in exactly one of the selected members of N, or

(c) in at least one of the selected members of N.

Let $x_j = 1(0)$ if N_j is (is not) selected, and let $a_{ij} = 1(0)$ if $i \in N_j$ ($i \notin N_j$), $i = 1, ..., m, j = 1, ..., n$. By 1_m we denote the m-dimensional vector of ones, i.e., $1_m = (1, 1, ..., 1)$.

The (weighted) *set packing problem* corresponds to case (a) and may be formulated as

$$v(SP) = \max \sum_{j=1}^{n} c_j x_j, \tag{SP}$$

subject to

$$Ax \leqslant 1_m,$$
$$x_j = 0 \text{ or } 1, \quad j = 1, \ldots, n.$$

Case (b) describes the (weighted) *set partitioning problem*

$$v(SPP) = \max \{ cx | Ax = 1_m, \, x \in \{0, 1\}^n \}, \tag{SPP}$$

while case (c) corresponds to the (weighted) *set covering problem*

$$v(SC) = \min \{ cx | Ax \geqslant 1_m, \, x \in \{0, 1\}^n \}. \tag{SC}$$

If $c = 1_n$, i.e., if the weights c_j are all equal to one, one is interested in packing M in the maximal number of acceptable subsets, or in partitioning M in the maximal number of such subsets, or in covering M by the minimal number of such subsets.

The most widely used application seems to be the airline crew scheduling problem, in which M corresponds to the set of flight legs to be covered during a planning period (usually, a week or two weeks after which the flight-schedule repeats itself), while N_j stands for a possible rotation (sequence of flight-legs with the same initial and terminal point) for a crew. In order to be acceptable, a rotation must satisfy certain regulations on the duration of the working time of a crew, time for rest, etc. Let c_j be a cost of a rotation N_j, $j = 1, \ldots, n$. Then the airline crew scheduling problem may be formulated as the set partitioning problem with obvious changes of the operator max for min. Some formulations of the airline crew scheduling problem permit "deadheading" of crews, i.e., the assignment of more than one crew to a flight-leg. In this case, the problem is treated as a set covering problem.

In a similar way, one may formulate railroad crew scheduling, truck deliveries, information retrieval and the facility location problem considered in Section 1.5 (see Section 8.6 for references).

8.1.1. *Basic Properties*

To rule out trivial pathologies, we assume that $A = (a_{ij})$, $i = 1, \ldots, m$, $j = 1, \ldots, n$, has neither a zero column nor a zero row. Note that these assumptions imply, e.g., that $F(SP) \neq \emptyset$ and $F(SC) \neq \emptyset$, while the feasibility of SPP is a far more complicated matter. Obviously, $F(SPP) \subseteq F(SP)$ and $F(SPP) \subseteq F(SC)$.

We show now that SPP can be brought to both the form of SP and SC while ensuring that the respective optimal solutions coincide.

THEOREM 8.1. *If $x^* \in F^*(SPP)$, then x^* is an optimal solution to the set covering problem, with the same matrix A in which the vector c is changed to c', with $c' = \alpha 1_m A - c$, where α is a sufficiently large number, e.g.,*

$$\alpha > \sum_{j=1}^{n} c_j. \tag{8.1}$$

Proof. The set partitioning problem may be written as

$$v(SPP) = \min \{ -cx + \alpha 1_m y | Ax - y = 1_m, \, y \geqslant 0, \, x \in \{0, 1\}^n \}.$$

Since $y = Ax - 1_m$, we have

$$v(SPP) = -m\alpha + \min\{cx'|Ax \geqslant 1_m, x \in \{0, 1\}^n\},$$

where $c' = \alpha 1_m A - c$. Thus, the above problem has the form of a set covering problem. $\qquad \square$

In a similar way, we prove

THEOREM 8.2. *If* $x^* \in F^*(SPP)$, *then* x^* *is an optimal solution to the set packing problem with the same matrix* A *in which the vector* c *is changed to* $c'' = c + \alpha 1_m A$, *where* α *satisfies* (8.1). $\qquad \square$

8.1.2. *A Transformation of the Knapsack Problem into the Set Packing Problem*

Now we construct for a given knapsack problem

$$v(K) = \max\{cx|ax \leqslant b, x \in \{0, 1\}^n\}, \tag{K}$$

an equivalent set packing problem. We know (Section 5.1) that without loss of generality we may assume that all data in K are positive integers and $a_j \leqslant b$ for $j = 1, ..., n$. First we replace variable x_j by a_j new variables x_{jk} for $k = 1, ..., a_j$, $j = 1, ..., n$. Replacing each term $a_j x_j$ by the expression

$$\sum_{k=1}^{a_j} x_{jk}, \tag{8.2}$$

we must ensure that all variables x_{jk} are either equal to zero simultaneously or, alternatively, equal to one simultaneously. To do this, we introduce a new binary variable (indicator) z_j for $j = 1, ..., n$ and require that $x_{jk} + z_j = 1$ for $k = 1, ..., a_j$, $j = 1, ..., n$. Furthermore, in order to compute the objective function value correctly, we define coefficients $c_{jk} = c_j/a_j$ for each variable x_{jk}. Then K is equivalent to

$$v(K_1) = \max \sum_{j=1}^{n} \sum_{k=1}^{a_j} c_{jk} x_{jk}, \tag{K_1}$$

subject to

$$\sum_{j=1}^{n} \sum_{k=1}^{a_j} x_{jk} \leqslant b, \tag{8.3}$$

$$x_{jk} + z_j = 1, \quad j = 1, ..., n, k = 1, ..., a_j,$$

$$x_{jk} = 0 \text{ or } 1 \quad \text{and} \quad z_j = 0 \text{ or } 1 \quad \text{for all } j \text{ and } k.$$

In order to obtain a set packing problem from K_1, we introduce for all j and k new variables x_{jkl}, with objective function coefficients $c_{jkl} = c_{jk}$ for $l = 1, ..., b$. (see (8.2)). Now we can write down the equivalent problem K_2 of the form

$$v(K_2) = \max \sum_{j=1}^{n} \sum_{k=1}^{a_j} \sum_{l=1}^{b} c_{jkl} x_{jkl}, \tag{K_2}$$

subject to

$$\sum_{j=1}^{n}\sum_{k=1}^{a_j} x_{jkl} \leqslant 1 \quad \text{for} \quad l = 1, \ldots, b,$$

$$\sum_{l=1}^{b} x_{jkl} + z_j = 1 \quad \text{for} \quad j = 1, \ldots, n, k = 1, \ldots, a_j, \tag{8.4}$$

$$x_{jkl} = 0 \text{ or } 1 \quad \text{and} \quad z_j = 0 \text{ or all } j, k \text{ and } l.$$

Constraints (8.4) may be transformed into the set packing form in the same way as in the proof of Theorem 8.1. So, finally, we obtain a set packing problem equivalent to the knapsack problem.

The above considerations may be generalized for the case of the multidimensional knapsack problem (Exercise 8.7.2). In a similar way, one may prove the same relations between the simple plant location problem and the set covering problem.

8.1.3. Reductions

Similarly as in the case of the knapsack problem, it is often possible, a priori, to eliminate certain rows and columns from the matrix A in a given set packing, partitioning or covering problem. In other words, it is often possible to determine optimal values for certain variables. We will consider here only the reduction rules for the set covering problem and the set partitioning problem with the obvious change of the operator from max to min, while in Exercise 8.7.4, we ask for similar reduction rules for the set packing problem.

Let a_i (a_j) denote the ith row (the jth column) of matrix A. The rules below are applied sequentially to matrix A.

1. If a_i is a null vector for some i, then $F(SC) = \emptyset$ and $F(SPP) = \emptyset$.

2. If $a_i = e_k$ (the kth unit vector) for some i, k, then $x_k^* = 1$ in every feasible solution to SC and SPP and the ith constraint may be deleted. Also, every row t such that $t \in N_k$ may be deleted. Additionally, for the set partitioning problem, every column $q \neq k$ such that $a_{tq} = a_{tk} = 1$, for some t, must be deleted as otherwise

$$\sum_{j=1}^{n} a_{tj} x_j \geqslant 2.$$

3. If $a_r \geqslant a_i$, then the rth constraint may be deleted as every cover of a_i covers a_r. In addition to deleting a_r in the set partitioning problem, we must delete every column k, i.e., set $x_k^* = 0$ such that $a_{rk} = 1$ and $a_{ik} = 0$, because, otherwise, if we set $x_k^* = 1$, then

$$\sum_{j=1}^{n} a_{rj} x_j \geqslant 2.$$

4. If for some set of columns S and some column k,

$$\sum_{j \in S} a_j = a_k \quad \text{and} \quad \sum_{j \in S} c_j \leqslant c_k,$$

then column k must be deleted. For the set covering problem, the above relations may be written is the following way: if

$$\sum_{j \in S} a_j > a_k \quad \text{and} \quad \sum_{j \in S} c_j \leqslant c_k,$$

then $x_k^* = 0$ is an optimal solution to SC.

8.2. OPTIMIZATION PROBLEMS OVER GRAPHS

In Chapter 3, we show that some linear programming problems may be formulated over appropriately constructed graphs (networks). In this section, we demonstrate that such an approach is useful in the case of packing and covering problems.

To transform the set packing problem of the form

$$v(SP) = \max\{cx|Ax \leqslant 1, x \in \{0, 1\}^n\} \tag{SP}$$

into an optimization problem over a finite undirected graph, we associate with A the *intersection graph* $G_A = (V, E)$ in the following manner. G_A has a vertex (node) for every column A and an edge connecting nodes j and k if and only if column k and column j of A satisfy $a_{ij} = a_{ik} = 1$ for at least one row $i \in \{1, ..., m\}$, i.e., when the intersection of N_j and N_k is nonempty. Let A_G be the edges versus nodes incidence matrix of the graph G_A and associate with node j of G_A the weight c_j for $j = 1, ..., n$, and suppose $|E| = q$. Then the (weighted) *node packing* problem may be formulated as

$$v(NP) = \max\{cx|A_G x \leqslant 1_q, x \in \{0, 1\}^n\}. \tag{NP}$$

It is easy to see that there exists a one-to-one correspondence between $x \in F(NP)$ and $x \in F(SP)$. So we have proved

THEOREM 8.3. *Problem NP is equivalent to problem SP.*   □

In other words, any set packing problem is equivalent to a node packing problem on a finite undirected graph with n nodes.

By definition, the matrix A_G has exactly two ones per row. Thus, by making the substitution $z_j = 1 - x_j$, we obtain the so-called *node covering problem*

$$v(NC) = \min\{cx|A_G x \geqslant 1_q, x \in \{0, 1\}^n\}. \tag{NC}$$

We note that every $x \in F(NC)$, i.e., every node-cover of a general graph G, corresponds uniquely to an independent set of nodes in G_A, i.e., to $x \in F(NP)$, and vice versa. Furthermore, whereas a set packing problem in n variables is always equivalent to a node packing problem in some graph G on n nodes, a set covering problem in n variables is, generally, not equivalent to a node covering problem in a graph on the same number of nodes.

Let now $G = (V, E)$ be a general finite undirected graph with $|V| = n$ and $|E| = q$. Let A_G be the edges (rows of A_G) versus nodes (columns of A_G) incidence matrix of a graph G. Replacing the word "node" by "edge" and vice versa, we obtain the so-called *edge packing problem*

$$v(EP) = \max\{cx|A_G^T x \leqslant 1_n, x \in \{0, 1\}^q\}, \qquad (EP)$$

where A_G^T denotes the transposition of A_G and c_j is now the weight assigned to edge $e_j \in E$, $j = 1, ..., q$.

The edge packing problem is often called the *edge matching problem* as in this problem we are looking for a subset of edges of some graph G such that no two edges have a node of G in common, while the edge-weight is maximum. If $c_j = 1$ for $j = 1, ..., q$, we are looking for a subset of edges of maximal cardinality and EP is sometimes called the maximal edge matching problem. If in the maximal edge matching problem we have $A_G^T x = 1_n$, we say that G has a *perfect-matching* as then every node of G is incident to an edge from that subset, or in other words, then G may be partitioned into mutually non-adjacent edges. It is easy to see that EP is a particular case of the set packing problem since in EP every row of A_G has exactly two ones. Similarly, the problem of finding a perfect matching in a given graph G is a particular case of the set partitioning problem.

Replacing the operator max by min and reversing the inequality, we obtain the so-called *edge covering problem*

$$v(EC) = \min\{cx|A_G^T x \geqslant 1_n, x \in \{0, 1\}^q\}, \qquad (EC)$$

which is a special case of the set covering problem.

8.3. PACKING AND COVERING POLYTOPES

Similarly as in the case of the knapsack problem (Section 5.5), we define the (set) packing polytope as the convex hull of all feasible solutions to SP, i.e., $\text{conv}\,F(SP)$. By Theorem 8.3, $\text{conv}\,F(SP) = \text{conv}\,F(NP)$. We define the feasible region of the linear programming relaxation of SP in the usual way:

$$F(\overline{SP}) = \{x \in R^n | Ax \leqslant 1_m, 0 \leqslant x \leqslant 1_n\}.$$

Note that constraints $x \leqslant 1_n$ are redundant in the definition of $F(\overline{SP})$. It is interesting both from the theoretical and practical points of view to find conditions under which $\text{conv}\,F(SP) = F(\overline{SP})$. Whenever $F(\overline{SP}) = \text{conv}\,F(SP)$, the set packing problem may be solved as a linear programming problem.

In a similar way, we define the *covering polytope* $\text{conv}\,F(SC)$ and ask when $\text{conv}\,F(SC) = F(\overline{SC})$, where

$$F(\overline{SC}) = \{x \in R^n | Ax \geqslant 1_m, 0 \leqslant x \leqslant 1_n\}.$$

A graph $G = (V, E)$ is called *bipartite* if V may be partitioned into two nonempty sets V_1 and V_2 and $E \subseteq V_1 \times V_2$. It may be proved that G is bipartite if and only if it has no odd cycles, i.e., cycles containing odd numbers of edges (Exercise 8.7.7).

THEOREM 8.4. *Let G be a graph having n nodes and q edges and let A_G be the $q \times n$ edge-node incidence matrix of G. Then the following three statements are equivalent:*

(i) *G is a bipartite graph,*

(ii) $F(\overline{SP}) = \operatorname{conv} F(SP),$

(iii) $F(\overline{SC}) = \operatorname{conv} F(SC).$

Proof. ((i) $\Rightarrow$ (ii)). If G is bipartite, then A_G is totally unimodular and thus (ii) follows by Theorem 3.4, i.e., every extreme point of $F(SP)$ is an integer.

((ii) $\Leftrightarrow$ (iii)). To show that (ii) and (iii) are equivalent, we note that the transformation $z_j = 1 - x_j$ transforms every extreme point of $F(\overline{SC})$ into an extreme point of $F(\overline{SC})$. On the other hand, the same transformation transforms an extreme point $\bar{x} \in F(\overline{SP})$ with a maximal number of components equal to one into an extreme point of $F(\overline{SC})$. Thus $F(\overline{SP})$ has fractional extreme points if and only if $F(\overline{SC})$ has a fractional extreme point.

To finish the proof, we assume that (i) does not hold. Then G contains an odd cycle on nodes $1, \ldots, 2k+1$, say, where $k \leqslant \lfloor n/2 \rfloor$. Then the vector $\bar{x}$ with components $\bar{x}_j = 1/2$ for $j = 1, \ldots, 2k+1$ and $\bar{x}_j = 0$ for $j \geqslant 2k+2$ is a fractional extreme point of $F(\overline{SP})$, since the edge-node incidence matrix of an odd cycle is nonsingular. Thus (ii) cannot hold.   □

For the edge packing problem EP and the edge covering problem EC, only a part of Theorem 8.4 is true.

THEOREM 8.5. *Under the same assumptions as in Theorem 8.4, the following two statements are equivalent:*

(i) *G is a bipartite graph,*

(ii) $F(\overline{EP}) = \operatorname{conv} F(EP).$ □

Usually, $F(\overline{SP}) \neq \operatorname{conv} F(SP)$, and it is necessary to add additional cuts to have $F(\overline{SP}) = \operatorname{conv} F(SP)$. Up to now, we know all of the cuts only for the case of the edge packing problem.

Let us consider a general graph $G = (V, E)$ with $|V| = n$ and $|E| = q$. In the proof of Theorem 8.4, we showed that a fractional extreme point corresponds to an odd cycle. Additional cuts therefore have to eliminate such cycles. Let $S \subseteq V$ be a set of nodes of such a cycle, so $|S| = 2k+1$, $k = 1, 2, \ldots \lfloor (n-1)/2 \rfloor$. Let $E(S)$ be a set of edges of G of a given cycle. The cut

$$\sum_{j \in E(S)} x_j \leqslant \tfrac{1}{2}(|S| - 1) \tag{8.5}$$

added to EP eliminates this cycle. Moreover, it may be proved that cuts (8.5) together with $F(\overline{EP})$ give $\operatorname{conv} F(EP)$.

THEOREM 8.6. *Let $G = (V, E)$ be a graph with $|V| = n$ and $|E| = q$. Then*

$$\operatorname{conv} F(EP) = F(\overline{EP}) \cap \left\{ x \in R^q \,\middle|\, \sum_{j \in E(S)} x_j \leqslant \tfrac{1}{2}(|S| - 1), S \in \overline{S} \right\},$$

where $\overline{S}$ is a family of all odd cycles of G.   □

For all of the packing problems considered so far, we may introduce a notion of a minimal cover and, by an analogy of the lifting procedure (Section 5.5), generate facets of $\operatorname{conv} F(SP)$ in a similar way as in the case of the knapsack problem.

By Theorem 8.3, the set packing problem is equivalent to the node packing problem defined on the corresponding intersection graph $G_A = (V, E)$. Now we may ask the question: which subgraphs of G_A "produce" facets of $\operatorname{conv} F(SP)$?

Let $G = (V, E)$ be a graph with n nodes and q edges and let A_G be its edge versus nodes incidence matrix.

THEOREM 8.7. *The inequality*

$$\sum_{j \in K} x_j \leqslant 1, \tag{8.6}$$

with $K \subset V$, is a facet of $\operatorname{conv} F(NP)$ *if and only if K is a clique of G, i.e., a maximal complete subgraph of G.* $\qquad\square$

We note that cuts (8.6) have the same form as constraints in a set packing problem, while cuts (8.5) do not have such a property.

8.4. THE TRAVELLING SALESMAN PROBLEM

In Section 1.5, we formulated the travelling salesman problem (*TSP*) as a problem of finding a sequence of visiting n cities. Let $C' = (c'_{ij})$, $i = 1, \ldots, n$, $j = 1, \ldots, n$, be the matrix of distances between these cities. Obviously, $c'_{ij} \geqslant 0$ and we set $c'_{ij} = \infty$ if city i is not directly connected with city j.

The problem may be formulated on an appropriately chosen graph in an obvious way: there exists an edge between node (city) i and j if and only if $c'_{ij} < \infty$. Let $G = (V, E)$ be such a graph. If for at least one pair (i, j) we have $c'_{ij} \neq c'_{ji}$, then G is a directed graph and we have an asymmetric *TSP*, otherwise, we have a symmetric *TSP*. A shortest closed path passing through every node of G is an optimal solution to the *TSP*. A simple numerical example demonstrates that, in general, such a path passes through some nodes of G more than one time. It can be shown that if basing on C' we compute the matrix $C = (c_{ij})$, $i = 1, \ldots, n, j = 1, \ldots, n$, of the shortest distances between each pair of nodes (i, j), then an optimal solution to the *TSP* passes through each node exactly once. Obviously, the elements of C satisfy the triangle inequality $c_{ij} + c_{jk} \geqslant c_{ik}$ for all i, j and k.

It should be noted that we do not always transform C' into C. For instance, in graph theory it is interesting to know whether a given graph $G = (V, E)$ is Hamiltonian, i.e., whether it admits a *Hamiltonian tour* or a tour which passes exactly once through every node of G. We answer this question if we set $c_{ij} = 1$ for all $(i, j) \in E$ and $c_{ij} = +\infty$ if $(i, j) \notin E$ and solve the corresponding *TSP*.

From Section 1.5 w know that the *TSP* may be formulated as

$$v(TSP) = \min \sum_{i=1}^{n} \sum_{j=1}^{n} c_{ij} x_{ij}, \tag{8.7}$$

subject to

$$\sum_{j=1}^{n} x_{ij} = 1, \quad i = 1, \dots, n, \tag{8.8}$$

$$\sum_{i=1}^{n} x_{ij} = 1, \quad j = 1, \dots, n, \tag{8.9}$$

$$\sum_{i \in S} \sum_{j \in V-S} x_{ij} \geq 1 \quad \text{for all } S \subset V, \ S \neq \varnothing \text{ and } S \neq V, \tag{8.10}$$

$$x_{ij} = 0 \text{ or } 1, \quad i = 1, \dots, n, \ j = 1, \dots, n. \tag{8.11}$$

The problem (8.7)–(8.9) and (8.11) is called the *assignment problem* (see Section 1.5) and we will denote it by *AP*. It is easy to see that every feasible solution to *AP* is a permutation of a set $V = \{1, \dots, n\}$ so $|F(AP)| = n!$ Similarly, every feasible solution to *TSP* is a cyclic permutation of the set of nodes of the graph $G = (V, E)$. Therefore, $|F(TSP)| = (n-1)!$

The constraint matrix of the assignment problem is totally unimodular (see Section 3.4). Thus, the problem may be solved by the simplex method replacing (8.11) by $x_{ij} \geq 0$ for all i, j, but due to the specific structure of the constraints, it is possible to construct more efficient methods, among them the so-called *Hungarian method* is probably most popular. We will discuss it in the next section.

Let φ be a permutation of a given set $N = \{1, \dots, n\}$ and let P_n be the set of all such permutations. Then the assignment problem may be formulated as

$$v(AP) = \min_{\varphi \in P_n} \sum_{i=1}^{n} c_{i\varphi(i)}. \tag{AP}$$

Similarly, if T_n is the set of all cyclic permutations of $N = \{1, \dots, n\}$, then

$$v(TSP) = \min_{\varphi \in T_n} \sum_{i=1}^{n} c_{i\varphi(i)}. \tag{TSP}$$

In Section 1.5, we define the *minimax assignment problem*, sometimes called the *bottleneck assignment problem*, as

$$v(MMAP) = \min_{\varphi \in P_n} \max_{i \in N} c_{i\varphi(i)}. \tag{MMAP}$$

In a similar way, we may define the *bottleneck travelling salesman problem*

$$v(BTSP) = \min_{\varphi \in T_n} \max_{i \in N} c_{i\varphi(i)}. \tag{BTSP}$$

Many practical problems connected with delivery of, e.g., goods from ware-houses to shops, and collecting, e.g., letters from post boxes, may be formulated in the form of the so-called *vehicle routing problem* with one depot. In this problem, $m > 1$ travelling salesmen have to visit n cities and go back to a given depot which, without loss of generality, may be chosen as city 1. The obvious generalization of the problem is the *vehicle routing problem with many depots*.

Now we show that the vehicle routing problem with one depot may be transformed into an equivalent *TSP*. Let $C = (c_{ij})$, $i, j = 1, \ldots, n$, be a given distance matrix for the classical *TSP*. Consider the matrix $\overline{C}$ obtained from C in the following way: above matrix C we write $m-1$ times the first row of C, and next $m-1$ times we repeat the first column of C (see Fig. 8.1). All elements of C located on

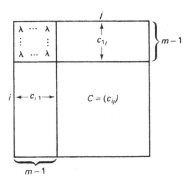

Fig. 8.1.

intersections of added rows and columns have a value equal λ. So we have m dummy cities which coincide with city 1. For different λ, we will have different problems:

1. $\lambda = \infty$. This means that the travelling salesman cannot travel between the dummy cities. Solving the classical *TSP* we obtain m disjoint cycles passing through city 1. This is equivalent to solving the vehicle routing problem with one depot with a given number of vehicles (travelling salesmen) equal to m.

2. $\lambda = -\infty$, which means that the travelling salesman may travel between dummy cities and moreover he gains some profit if he travels between such cities. Then, assuming that m is sufficiently large, the solution to *TSP* with the matrix $\overline{C}$ gives the minimal number of cycles passing through city 1, which is equivalent to the solution of the vehicle routing problem with one depot in which the number of vehicles is minimized.

3. $\lambda = 0$. Then the travelling salesman is neither penalized nor rewarded for his passage between dummy cities and the solution to *TSP* with $\overline{C}$ coincides with the solution obtained for the matrix C.

8.5. METHODS FOR *TSP*

Consider the *AP* with the matrix $C = (c_{ij})$, $i, j = 1, \ldots, n$, where all $c_{ij} \geqslant 0$. Subtracting from all elements of the ith row the smallest element α_i of this row, we obtain at least one zero in the ith row, $i = 1, \ldots, m$. We may repeat this for the jth column, $j = 1, \ldots, n$. If β_j is the smallest element of the jth column, $j = 1, \ldots, n$, one can easily show that a transformation of C into $H = (h_{ij})$ given by

$$h_{ij} = c_{ij} - \alpha_i - \beta_j \tag{8.12}$$

does not change an optimal solution to AP and moreover

$$v(AP) \geqslant \sum_{i=1}^{n} \alpha_i + \sum_{j=1}^{n} \beta_j. \tag{8.13}$$

As $v(TSP) \leqslant v(AP)$, (8.13) gives an upper bound on $v(TSP)$. A better bound is given by a solution of AP. One of the best known methods for AP is the Hungarian method, which in fact solves the dual to AP and is based on the following theorem of König and Egervery: m elements of C are called *independent* if no two of them are in the same row or column. A subset of rows and columns is called a 0-*covering* of C if all zero elements are only in a row or column of this subset.

THEOREM 8.8 (König, Egervary). *The maximal number of independent* 0-*elements of a given matrix* $C = (c_{ij})$, $i, j = 1, \ldots, n$, *is equal the minimal number of rows and columns of its* 0-*covering.*   □

If an optimal solution to the AP does not contain subcycles (i.e., cycles having less than n edges), it is at the same time an optimal solution to the TSP. Assume that this is not the case and let $S_1, S_2, \ldots, S_p$ be the edge set of subcycles in a given optimal solution to AP. To remove these subcycles, we add to the AP the constraint

$$\sum_{(i, j) \in S_k} x_{ij} \leqslant |S_k| - 1, \tag{8.14}$$

where S_k is one of subcycles of $S_1, S_2, \ldots, S_p$ and solve again the extended AP. Choosing S_k we may take into account the rules discussed in Section 4.3. The constraint (8.14) is satisfied if we choose $x_{i*j*} = 0$, and $x_{ij} = 1$ otherwise. This is equivalent to setting $c_{i*j*} = \infty$. The above method is called the *subcycles elimination method* and is particularly efficient for asymmetric TSP's.

Symmetric TSP's are difficult to solve by the above approach since the solutions of the corresponding AP's have in general many subcycles of length two. Much better bounds may be computed using Lagrangean relaxation of the TSP. Observe that (8.10) eliminates subcycles in a given symmetric graph $G = (V, E)$, with $V = \{v_1, \ldots, v_n\}$. A 1-tree of G is a graph (V, T) consisting of a tree on the nodes $\{v_2, \ldots, v_n\}$, together with two edges incident with vertex v_1. The Lagrangean relaxation is based on the following three observations:

(1) a Hamiltonian cycle is a 1-tree in which each node has a degree 2;

(2) a minimum 1-tree can be easily computed by a minimum spanning tree and the two edges incident with node v_1 which have minimal length;

(3) transformation (8.12) leaves the relative order of the solutions of the TSP invariant, but changes the minimum 1-tree.

The Lagrangean relaxation of (8.7)–(8.11) takes the form

$$v(L) = \min\left\{\sum_{i=1}^{n} \sum_{j=1}^{n} c_{ij} x_{ij} + \sum_{i=1}^{n} u_i \left(\sum_{j=1}^{n} x_{ij} - 1\right) + \sum_{j=1}^{n} v_j \left(\sum_{i=1}^{n} x_{ij} - 1\right)\right\},$$

subject to (8.10) and (8.11). It may be written as

$$v(L) = -\sum_{i=1}^{n} u_i - \sum_{j=1}^{n} v_j + \min \sum_{i=1}^{n} \sum_{j=1}^{n} (c_{ij} + u_i + v_j) x_{ij},$$

with restrictions (8.10) and (8.11).

For given u_i and v_j, solving $v(L)$ is easy since the coefficients of the objective function have the form of (8.12) and solving the shortest tree problem the constraint (8.10) and (8.11) are always satisfied. A method for computing the shortest tree will be discussed in the next chapter. Computational experiments show that $v(L)$ is a relatively sharp lower bound on $v(TSP)$. Combining subgradient methods for computing optimal or suboptimal values for u_i and v_j with a branch-and-bound method, gives an efficient approach to solving symmetric TSP's.

8.6. BIBLIOGRAPHIC NOTES

8.1–8.3. Set packing and set partitioning problems are considered in the review papers by Balas and Padberg (1976) and by Padberg (1979). Garfinkel and Nemhauser (1973) give the transformation of the knapsack problem into the set packing problem. Padberg (1979), Nemhauser and Trotter (1974) and Wolsey (1976) study the set packing polytope. Edmonds (1965a, b) gives the complete description of the edge packing polytope (see also Pulleyblank, 1983). Krarup and Pruzan (1983) transform the simple plant location problem into the set covering problem. Marsten (1974) describes an efficient algorithm for the set partitioning problem, while Etcheberry (1977) uses the Lagrangean relaxations in a branch-and-bound method for the set covering problem.

8.4 and 8.5. The basic references are here papers by Bellmore and Nemhauser (1968) and by Burkard (1979). The Hungarian method is described in Kuhn (1955). Generalizations of the assignment problem are described in Burkard et al. (1977), while the algorithms for the bottleneck assignment problem are given by Derigs and Zimmerman (1978) and Słomiński (1976). Methods for the vehicle routing problems are considered by Christofides (1976a) and Clarke and Wright (1964). Dantzig et al. (1954) propose the subcycle elimination method which was improved by Christofides (1972) and by Balas and Christofides (1981) (see also Bellmore and Malone, 1971). Held and Karp in 1970 and 1971 proposed using the Lagrangean relaxation for solving TSP. Their idea was developed further by Hansen and Krarup (1974) and by Bazaraa and Goode (1977). Miliotis (1978) uses the cutting plane method in solving TSP.

8.7. EXERCISES

8.7.1. Prove Theorem 8.2.

8.7.2. Demonstrate that the multidimensional knapsack problem is equivalent to the set packing problem.

8.7.3. Transform the simple plant location problem into an equivalent set covering problem.

8.7.4. Construct the reduction rules for the set packing problem.

8.7.5. For $c = (7, 5, 8, 6, 10, 13, 11, 4, 8)$ and

$$A = \begin{bmatrix} 1 & 1 & 1 & 0 & 1 & 0 & 1 & 1 & 0 \\ 0 & 1 & 1 & 0 & 0 & 0 & 0 & 1 & 0 \\ 0 & 1 & 0 & 0 & 1 & 1 & 0 & 1 & 1 \\ 0 & 0 & 0 & 1 & 0 & 0 & 0 & 0 & 0 \\ 1 & 0 & 1 & 0 & 1 & 1 & 0 & 0 & 1 \\ 0 & 1 & 1 & 0 & 0 & 0 & 1 & 0 & 1 \\ 1 & 0 & 0 & 1 & 1 & 0 & 0 & 1 & 1 \end{bmatrix}$$

write the set packing, covering and partitioning problem, reduce them and next solve them.

8.7.6. Write the node packing problem for the data from Exercise 8.7.5 and solve it.

8.7.7. Demonstrate that a graph is bipartite if and only if it contains no odd cycles.

8.7.8. Prove that the incidence matrix of an odd cycle is nonsingular and the corresponding basic solution has the form $(1/2, 1/2, ..., 1/2)$.

8.7.9. Prove Theorem 8.5.

8.7.10. Prove Theorem 8.6.

8.7.11. Construct the graph G_4 for the data from Exercise 8.7.5 and next solve the edge packing problem with $c = (1, 1, ..., 1)$.

8.7.12. Write all additional cuts for the problem from Exercise 8.7.11 using Theorem 8.6.

8.7.13. Prove Theorem 8.7.

8.7.14. Write a branch-and-bound algorithm for the set packing problem based on reduction and constraint generation.

8.7.15. Show that transformation (8.12) does not change $x^* \in F^*(AP)$ and $x^* \in F^*(TSP)$.

8.7.16. Write the Hungarian method for the AP.

8.7.17. Write the branch-and-bound method for the asymmetric TSP (Balas and Christofides, 1976).

CHAPTER 9

Near-Optimal Methods

We begin with a classification of near-optimal methods and a discussion of three ways of evaluating them. Next, in Section 9.2, we consider the greedy algorithm. The so-called approximation algorithms may be considered as extensions of the greedy algorithm as they provide a feasible solution whose suitable defined error is not greater than a given number. In Section 9.4, we consider near-optimal methods for solving the travelling salesman problem. This problem is an example of problems for which the approximation algorithms do not exist. At the end of this chapter, we again consider the greedy algorithm using the notions of independence systems and matroids.

9.1. CLASSIFICATION AND EVALUATION OF NEAR-OPTIMAL METHODS

Generally speaking, a vector $x^* \in R^n$ is a *(global) optimal solution* of the problem

$$v(P) = \max_{x \in F(P)} f(x) \qquad\qquad (P)$$

if two conditions are satisfied:

(i) $x^* \in F(P)$,

(ii) $f(x^*) \geqslant f(x)$ for all $x \in F(P)$.

Usually, a *near-optimal solution* to P is a vector $\hat{x} \in R$ such that it satisfies the first condition, i.e., $\hat{x} \in F(P)$, and for which $f(\hat{x})$ is "close" to $f(x^*)$, although, as we show in Section 7.1, in many practical problems, a near-optimal solution which is "close" to $F(P)$, in the sense of a suitably defined norm, has some value. For instance, in capital budgeting problems, some contraints may be considered soft ones, i.e., they may be violated by a near-optimal solution $\hat{x}$.

In such circumstances, after suitable modifications, any exact method may be considered a near-optimal one. For instance, we may always interrupt a branch-and-bound method and consider the best feasible solution found so far as a near-optimal one.

A relatively large class of near-optimal methods is constituted by methods which provide a *local optimum* to a given problem P, i.e., they find the best solution in a suitable defined neighbourhood of a given point $x' \in F(P)$. For instance, for a given integer problem

$$v(P) = \max \{cx | Ax \leqslant b, x \geqslant 0, x \in Z^n\}, \qquad\qquad (P)$$

the neighbourhood of x' may be defined as

$$N(x') = \{x \in R^n | x_j = x'_j - 1, x'_j, x'_j + 1, j = 1, \ldots, n\}.$$

Then $\hat{x} \in R^n$ is a *local maximum* to P in $N(x')$ if

$$c\hat{x} \geqslant cx \quad \text{for all } x \in N(x') \cap F(P).$$

A vector $x^* \in R^n$ is a global optimum if it is a local optimum in any neighbourhood of x^*.

For some problems, it is not necessary to consider all possible neighbourhoods of a given point x' in order to prove that $x' \in F^*(P)$. For instance, in linear programming, it is sufficient to define the neighbourhood of a vertex x' as a set of all vertices adjacent to x', i.e., the set of vectors which differ from x' in exactly one coordinate. Then we know from Section 2.1 that any local optimum coincides with a global one.

As there are many near-optimal methods for a given problem, it is natural to ask how such methods should be evaluated. In general, we distinguish three approaches:

(1) computational experiments,

(2) probabilistic analysis,

(3) worst case analysis.

In *computational experiments*, an algorithm is tried out on a set of supposedly representative test problems and the results of such experiments are evaluated statistically. For instance, we may study the difference between values of exact and near-optimal solutions for a given set of test problems or the difference between solution times for a given exact and near-optimal algorithm. The obvious drawback of such an approach is the difficulty in connecting with an extension of the evaluation for the problems not included in the set of test problems.

The *probabilistic analysis of algorithms* partially removes this drawback. This analysis starts from a specification of what an average problem instance would look like, in terms of a probability distribution over a class of all instances. The running time and the solution value of a particular algorithm are then random variables, whose behaviour can be studied and evaluated. So far this analysis was applied to relatively simple algorithms under the assumption that $n \to \infty$, where n is the number of variables.

The *worst case analysis* provides estimations for the maximal deviation from optimality defined in the following way.

We know (see Section 1.6) that a problem P is, in general, a infinite set of problem instances (numerical examples, tasks). For instance, for (integer) knapsack problem

$$v(K) = \max \{cx | ax \leqslant b, x \geqslant 0, x \in Z^n\}, \tag{K}$$

every $(2n+1)$-dimensional vector $(c_1, ..., c_n, a_1, ..., a_n, b)$ defines an instance of K, providing all its coordinates are positive integers such that $a_j \leqslant b$ for all j. In general, for a given instance (task) $t \in P$, by $v(t)$ we denote the value of this task, i.e., the value of the objective function at an optimal point. In this chapter, we will consider near-optimal methods which provide a feasible solution $x' \in R^n$ such that its value $h(t)$ is uniquely defined. Then, for maximization problems, $h(t) \leqslant v(t)$ for all $t \in P$, and in the worst case analysis we are looking for a largest possible real number $\delta \leqslant 1$ such that

$$h(t) \geqslant \delta v(t) \qquad \text{for all } t \in P \tag{9.1}$$

under the assumption that $v(t) > 0$ for all $t \in P$. In the case of minimization problems, we are seeking a smallest possible real number $\delta \geqslant 1$ such that

$$h(t) \leqslant \delta v(t) \qquad \text{for all } t \in P. \tag{9.2}$$

Again we assume that $v(t) > 0$ for all $t \in P$.

Knowing δ we define the maximal relative error for the maximization problems as

$$\varepsilon = \sup_{t \in P} \left(1 - \frac{h(t)}{v(t)}\right) = \sup_{t \in P} \frac{v(t) - h(t)}{v(t)}, \tag{9.3}$$

while for the minimization problems we have

$$\varepsilon = \inf_{t \in P} \frac{h(t) - v(t)}{v(t)}. \tag{9.4}$$

So δ and ε may be considered as some of the evaluations of a given near-optimal method.

The next problem studied in the worst case analysis is whether a given bound δ is obtainable, i.e., whether there exists a task $t^* \in P$ such that $h(t^*) = \delta v(t^*)$ or does a sequence of tasks $t \in P$ exist such that $h(t)/v(t)$ converges to δ.

Obviously, we will consider only polynomial algorithms as near-optimal methods and apply them to $\mathcal{NP}$-hard problems (see Section 1.6). Then an interesting question from the practical and theoretical point of view arises: how large may the relative error be if we solve an $\mathcal{NP}$-hard problem by a given polynomial algorithm? Similarly as in the case of exact methods, the most advanced results have been obtained for the case of the knapsack problem. Therefore, in the next two sections, we will consider near-optimal methods for the knapsack problem.

9.2. AN ANALYSIS OF THE GREEDY ALGORITHM

In Chapter 5, we defined the greedy algorithm for the (binary) knapsack problem. Here we modify it for the integer knapsack problem of the form

$$v(K) = \max \left\{ \sum_{j=1}^{n} c_j x_j \, \bigg| \, \sum_{j=1}^{n} a_j x_j \leqslant b, \, x \geqslant 0, \, x \in Z^n \right\}, \tag{K}$$

where all the data are positive integers and $a_j \leqslant b$ for all j. Without loss of generality, we assume that $0 \leqslant x_j \leqslant d_j$, where

$$d_j = \lfloor b/a_j \rfloor, \qquad j = 1, \dots, n.$$

Then, in the binary knapsack problem $d_j = 1$ for all j. We also assume (see Section 5.1) that

$$c_1/a_1 \geqslant c_2/a_2 \geqslant \dots \geqslant c_n/a_n. \tag{9.5}$$

The Greedy Algorithm

Step 0 (Initialization): $j := 1; \overline{b} := b; \hat{x} := 0;$

Step 1 (Main Iteration): $\hat{x}_j := \min\{d_j, \overline{b}/a_j\}; \overline{b} := \overline{b} - a_j\hat{x}_j;$

Step 2 (Stop): If $j = n$ or $\overline{b} = 0$, then stop $(h(t) := c\hat{x})$, otherwise, $j := j+1$ and go to Step 1.   □

To compute $\hat{x}$, the greedy algorithm makes $O(n)$ comparisons and ordering (9.5) requires $O(n\log n)$ comparisons.

THEOREM 9.1. *The greedy algorithm applied to the integer knapsack problem gives* $\varepsilon \leqslant 0.5$ *and this estimation is sharp.*

Proof. By our assumptions, $h(t) \geqslant c_1 \lfloor b/a_1 \rfloor$. As the value of the linear programming relaxation of K equals $c_1 b/a_1$, we have

$$\delta = \inf_{t\in K} \frac{h(t)}{v(t)} \geqslant \inf_{t\in K} \frac{\lfloor b/a_1 \rfloor}{b/a_1} = \frac{\lfloor b/a_1 \rfloor}{\lfloor b/a_1 \rfloor + (b/a_1 - \lfloor b/a_1 \rfloor)} \geqslant \frac{1}{2}.$$

Consider now a sequence of instances of the integer knapsack problem with the parameter $k = 1, 2, \ldots$, constructed in the following way for $n = 2$: $c_1 = a_1 = k+1, c_2 = a_2 = k, b = 2k$. Then $v(t) = 2k, h(t) = k+1$ and

$$\delta = \frac{h(t)}{v(t)} = \frac{k+1}{2k} \rightarrow \frac{1}{2}$$

when $k \rightarrow \infty$. Therefore, $\varepsilon = 1 - \delta = 0.5$ and this bound is sharp.   □

For the binary knapsack problem, ε is even larger since for a sequence of instances constructed for $n = 2$, with $c_1 = a_1 = 1, c_2 = a_2 = k$ and $b = k, k = 1, 2, \ldots$, we have $v(t) = k$, while the greedy algorithm gives $h(t) = 1$. Then

$$\delta \leqslant \frac{h(t)}{v(t)} = \frac{1}{h} \rightarrow 0$$

when $k \rightarrow \infty$. Therefore, $\varepsilon = 1 - \delta = 1$. We should point out that this is the worst case analysis and for randomly generated knapsack problems the greedy algorithm gives much better results. For instance, in computational experiments (see Section 9.6 for references), ε was not larger than 1%.

It is possible to give a theoretical justification why the greedy algorithm gives relatively good results for randomly generated instances. If for the integer knapsack problem we introduce the new parameter $q = \lfloor b/\max_{j\in N} a_j \rfloor$, where $N = \{1, \ldots, n\}$, then, by Theorem 9.1, we have better estimation of

$$\delta = \inf_{t\in K} \frac{h(t)}{v(t)} \geqslant \frac{\lfloor b/a_1 \rfloor}{b/a_1} \geqslant \frac{\lfloor b/a_1 \rfloor}{\lfloor b/a_1 \rfloor + 1} \geqslant \frac{q}{q+1}.$$

Similarly, for the binary knapsack problem, it is possible to prove

THEOREM 9.2. *The greedy algorithm applied to the binary knapsack problem gives*

$$\varepsilon = \sup_{t \in K} \frac{v(t)-h(t)}{v(t)} \leqslant 1 - \frac{1}{b}\sum_{j=1}^{p-1} a_j \leqslant 1 - \frac{p-1}{b},$$

where p is a fractional variable in an optimal solution to the linear programming relaxation of the problem.   □

9.3. APPROXIMATION METHODS

In approximation methods, we may decrease the relative error at the expense of their running time. For instance, below we show that an approximation algorithm with a complexity $O(n^2)$ gives a relative error $\varepsilon = 1/3$, while an algorithm with a complexity $O(n^3)$ has $\varepsilon = 1/4$, and so on. Moreover, we show that for any given $0 < r < 1$ an approximation algorithm with $\varepsilon \leqslant r$ may be constructed.

9.3.1. *A Polynomial Approximation Scheme*

An algorithm is called a *polynomial approximation scheme* for a maximization problem P if it solves in polynomial time any instance of P with the relative error $\varepsilon \leqslant r$, where r is a given number such that $0 < r < 1$.

According to the above definition, the greedy algorithm is not a polynomial approximation scheme as it is possible to construct a sequence of instances for which ε is as close to 1 as possible. But the greedy algorithm constitutes a part of a polynomial approximation scheme, which we will call Algorithm A6.

Let $h(S)$ be the value of the solution obtained by the greedy algorithm applied to the (binary) knapsack problem with variables from the set $\{1, \ldots, n\} - S$ and the right-hand side

$$b' = b - \sum_{j \in S} a_j.$$

Algorithm A6 solves the knapsack problem

$$h(t) = \max_{S} \left(\sum_{j \in S} c_j + h(S) \right),$$

subject to

$$\sum_{j \in S} a_j \leqslant b, \quad |S| \leqslant k,$$

i.e., Algorithm A6 considers all subsets S of $N = \{1, \ldots, n\}$ such that $|S| \leqslant k$, where k is a given integer, and next finds a greedy solution for the knapsack problem with variables $x_j, j \in N - S$.

THEOREM 9.3. *For given $0 < \varepsilon < 1$, Algorithm A6 is a polynomial approximation scheme if $k = \min\{n, \lceil 1/\varepsilon \rceil - 1\}$.*

Proof. If x^* is an optimal solution to K, define $S^* = \{j \in N | x_j^* = 1\}$. If $|S^*| \leqslant k$, obviously $h(t) = v(K)$. Otherwise, define S to be a set of k items of S^* with the largest profits and let p be an index of a fractional variable. Then

$$v(t) \leqslant \sum_{j \in S} c_j + \sum_{j=1}^{p-1} c_j + c_p \leqslant h(t) + c_p,$$

$$v(t) = \sum_{j \in S^*} c_j \geqslant (k+1) c_p.$$

Therefore,

$$v(t) - h(t) \leqslant c_p \leqslant v(t)/(k+1) \leqslant \varepsilon v(t).$$

So (9.3) holds. As the number of subsets of cardinality at most k can be bounded by n^{k+1}, the total complexity of Algorithm A6 is $O(kn^{k+1}) = O(kn^{1/\varepsilon})$. $\qquad\square$

If we want to find a near-optimal solution with guaranteed accuracy $\varepsilon = 20\%$, then $k = 1/\varepsilon - 1 = 4$, and we have to consider in Algorithm A6 all subsets S of N with $|S| \leqslant 4$.

9.3.2. *Fully Polynomial Approximation Schemes*

An algorithm is called a *fully polynomial approximation scheme* (*FAS*) if it is a polynomial approximation scheme and, additionally, it is a polynomial of $1/\varepsilon$. Therefore, Algorithm A6 is not a *FAS* as it is a polynomial of the size of the input, but it is an exponential of $1/\varepsilon$.

All *FAS*'s are based on the dynamic programming method considered in Chapter 5. We construct an *FAS* for the knapsack problem and next modify this construction to the case of a general integer programming problem. We know that the knapsack problem

$$v(K) = \max \{cx | ax \leqslant b, x \geqslant 0, x \in \{0, 1\}^n\} \qquad (K)$$

may be solved by dynamic programming using $O(nb)$ comparisons and additions. If

$$d = \sum_{j=1}^{n} c_j,$$

then one may easily prove that solving K by dynamic programming requires $O(nd)$ comparisons and additions (operations).

The idea of the *rounding method* is to solve by dynamic programming the knapsack problem with modified (rounded) profits $c_j' = \lceil c_j/q \rceil$, where q is a suitable choosen value.

Consider the knapsack problem K' with $c_j' = \lceil c_j/q \rceil$ for $j = 1, \ldots, n$. Then K' may be solved by dynamic programming using $O(nd/q)$ operations. Let $x' \in F^*(K')$. We estimate the maximal error made if we choose x' as a solution of K. For any instance $t \in K$, we have $h(t) = cx'$ and $v(t) \geqslant d/n$ since

$$\max_{1 \leqslant j \leqslant n} c_j \geqslant d/n$$

and $a_j \leqslant b$ for all j. As $q\lceil c_j/q\rceil \geqslant c_j$ for all j, x' is an optimal solution of K'' with $c_j = q\lceil c_j/q\rceil$ for all j. Then

$$v(t) \sum_{j=1}^{n} q\lceil c_j/q\rceil x'_j \leqslant \sum_{j=1}^{n} (c_j+q)x'_j \leqslant h(t)+nq.$$

Since $v(t) \geqslant d/n$, we have

$$h(t) \geqslant \left(1-\frac{nq}{v(t)}\right) v(t) \geqslant \left(1-\frac{n^2q}{d}\right)v(t).$$

Thus, for a given δ, we have to choose

$$q = \frac{(1-\delta)d}{n^2},$$

and then the computational complexity of the rounding method is

$$O(nd/q) = O\left(n^3/(1-\delta)\right) = O(n^3/\varepsilon).$$

So the rounding method is a *FAS* since the computational load grows polynomially with the size n and $1/\varepsilon$.

Consider now a general integer programming problem

$$v(P) = \max\left\{cx \mid Ax \leqslant b, x \geqslant 0, x \in \{0,1\}^n\right\}, \qquad (P)$$

where all data are nonnegative integers. Without loss of generality, we may assume that $a_{ij} \leqslant b_i$ for $i = 1, \ldots, m$ and $j = 1, \ldots, n$.

For $1 \leqslant k \leqslant n$ and $r_j = 0$ or $1, j = 1, \ldots, k$, a vector $(r_1, \ldots, r_k)$ is called a *partial solution* to P if there exists at least one $x \in F(P)$ such that $x_j = r_j$ for $j = 1, \ldots, k$. Let S_k be a set of all partial solutions for a given k. If P is solved by dynamic programming, then $|S_{k+1}| \leqslant 2|S_k|$ so the cardinality of S_k may grow exponentially. We show below that using rounding of the coefficients of the objective function of P and simple domination rules in S_k, it is possible to construct a *FAS* in which

$$\sum_{k=0}^{n} S_k \leqslant w(k, 1/\varepsilon),$$

where $w(\cdot)$ is some polynomial of k and $1/\varepsilon$, $S_0 = \{(0, \ldots, 0)\}$ and ε is a given guaranteed accuracy.

Let $\bar{c}$ be the largest coefficient in the objective function. We will use the same notation as in the case of the knapsack problem. By $\mathrm{rem}(a, b)$ we denote the remainder of the division of a by b. Let $c'_j = c_j - \mathrm{rem}(c_j, \bar{c}\varepsilon/n)$, $j = 1, \ldots, n$. For all $t \in P$ we have $\mathrm{rem}(c_j, \bar{c}\varepsilon/n) \leqslant \bar{c}\varepsilon/n$, therefore

$$\sum_{j=1}^{n} |c_j - c'_j| \leqslant \varepsilon\bar{c} \leqslant \varepsilon v(t).$$

To estimate the computational load, we denote by $c''_j = \lfloor c_j n/(\bar{c}\varepsilon)\rfloor$ for all j. Then $c''_j = c'_j n/(\bar{c}\varepsilon)$ for all j, and the sets S_n in problems P' and P'' have the same cardinality so the solution time for these problems is the same. Since $c'_j \leqslant \bar{c}$, therefore $c''_j \leqslant \lfloor n/\varepsilon \rfloor$ and thus

$$|S_k| \leqslant 1 + \sum_{j=1}^{k} c''_j \leqslant 1 + k \lfloor n/\varepsilon \rfloor,$$

$$\sum_{k=0}^{n-1} |S_k| \leqslant n + \sum_{k=0}^{n-1} k \lfloor n/\varepsilon \rfloor.$$

The last expression grows not faster than $O(n^3/\varepsilon)$. So we have constructed a *FAS* for P, called the *rounding method*.

The cardinality of S_k may be limited in another way. Let

$$V_k = \sum_{j=1}^{k} c_j r_j.$$

The idea of the *interval partitioning method* is to partition the interval $(0, V_k)$ into subintervals of the length $V_k/(n-1)$ (the last interval may be shorter) and to delete from each subinterval all partial solutions except one. Then the cardinality of all sets S_k may be bounded as $O(n^2/\varepsilon)$.

9.4. NEAR-OPTIMAL METHODS FOR *TSP*

The travelling salesman problem is an example of an $\mathcal{NP}$-hard integer optimization problem. Below we show that, in general, finding a near-optimal solution to *TSP* with a guaranteed accuracy ε (ε-solution) is an $\mathcal{NP}$-complete problem. Such an ε-solution has to satisfy (9.4).

Let $G' = (V, E')$ be a graph with $|V| = n$. After addition of missing edges, we have a complete graph $G = (V, E)$ with the cost of an edge defined as

$$c_{ij} = \begin{cases} 1 & \text{if} \quad (i,j) \in E', \\ k & \text{if} \quad (i,j) \notin E'. \end{cases}$$

Obviously, if $k > 1$, then $v(K) = n$ implies that there exists a Hamiltonian cycle in G'. If $k \geqslant (1+\varepsilon)n$, the unique ε-solution is a solution with $h(t) = n$. But this means that finding an ε-solution to *TSP* is as difficult as proving that a given graph has a Hamiltonian cycle which, as we know, is an $\mathcal{NP}$-complete problem. So we have proved

THEOREM 9.4. *For given $\varepsilon > 0$, finding an ε-solution is an $\mathcal{NP}$-complete problem.* $\square$

In other words, for any $\varepsilon > 0$, it is possible to construct an ε-approximation algorithm for *TSP* if and only if $\mathcal{P} = \mathcal{NP}$. There are many near-optimal methods for *TSP*, and for some of them under additional assumptions it is possible to estimate ε or δ.

One of the best near-optimal methods with $\delta = 3/2$ was proposed by Christofides. In this method, we additionally assume that the costs of edges satisfy the triangle inequality. Let $G = (V, E)$ with $|V| = n$ and $E = \{(i, j)|1 \leqslant i, j \leqslant n\}$. For any $S \subseteq E$, we define $v(S)$ as

$$v(S) = \sum_{(i,j) \in S} c_{ij}.$$

Let E_{ST} be a set of all edges from the shortest spanning tree for G, and let T be a set of all edges of an optimal Hamiltonian cycle. Obviously, $v(E_{ST}) \leqslant v(T)$ since after removing an edge we have a tree which may not be the shortest one.

By d_i we denote the *degree* of a node $i \in V$, i.e., the number of edges incident with i. Let V_0 be a set of all nodes with odd degrees in the shortest tree of G. Let $E_0 = \{(i, j) \in E | i \in V_0, j \in V_0\}$. It may be proved that $|V_0| = k$ is even and the graph $G_0 = (V_0, E_0)$ has a perfect matching E_{MP}. Without loss of generality, we may assume that $V_0 = \{1, \ldots, k\}$ and that $T_0 = \{(1, 2), (2, 3), \ldots, (k-1, k), (k, 1)\}$ is an optimal Hamiltonian cycle in G_0. Then

$$v(E_{PM}) \leqslant \tfrac{1}{2} v(T_0) \leqslant \tfrac{1}{2} v(T).$$

The first of the inequalities follows from the fact that $\{(1, 2), (3, 4), \ldots, (k-1, k)\}$ and $\{(2, 3), (4, 5), \ldots, (k, 1)\}$ are perfect machings in G_0. The second inequality follows from the triangle inequality in G.

Consider a graph $G' = (V, E_{ST} \cup E_{PM})$ which may have multiple edges. Since all modes in G' have even degrees, in G' there exists a closed path S passing through each node at least once and

$$v(S) = v(E_{ST}) + v(E_{PM}) \leqslant v(T). \tag{9.6}$$

Using the triangle inequality it is possible to modify S into a Hamiltonian cycle in G.

One of the simplest near-optimal methods for *TSP* is the so-called the *nearest-neighbour method* in which the travelling salesman goes from a given city i to the nearest city (neighbour). The method has complexity $O(n)$, but its error may be as big as possible.

Relatively good results have been obtained with the so-called *k-optimal method*. A Hamiltonian cycle T is k-optimal if exchange of its l edges into l edges from $E-T$ does not decrease the length of T for any $l \leqslant k$. Usually, for $k = 2$ or $k = 3$, one obtains good results.

9.5. INDEPENDENCE SYSTEMS AND MATROIDS

A feasible solution to the (binary) knapsack problem K has an interesting property. If $(x_1, \ldots, x_{k-1}, 1, x_{k+1}, \ldots, x_n) \in F(K)$, then $(x_1, \ldots, x_{k-1}, 0, x_{k+1}, \ldots, x_n)$ is also feasible. Basing on this observation we define the knapsack problem in a different setting.

Let E be a finite set and $\mathscr{F} \subseteq 2^E$ be a family of subsets of E satisfying two conditions:

(M1) $\emptyset \in \mathcal{F}$ (the empty set belongs to $\mathcal{F}$),

(M2) if $B \subseteq A \in \mathcal{F}$, then $B \in \mathcal{F}$.

Such a pair $S = (E, \mathcal{F})$ is called an *independence system*. A set $A \subseteq E$ is called *independent* if $A \in \mathcal{F}$, otherwise ($A \in 2^E - \mathcal{F}$) it is *dependent*. A maximal (with respect to set inclusion) independent set is called a *basis*. A set of all bases will be denoted by $\mathcal{B}$. A minimal dependent set is called a *circuit*. A set of all circuits of S is denoted by $\mathcal{C}$.

If for a given knapsack problem

$$v(K) = \max\{cx | ax \leqslant b, x \in \{0, 1\}^n\},\tag{K}$$

we take $E = \{1, ..., n\}$, $A = A(x) = \{j \in E | x_j = 1\}$,

$$w(A) = \sum_{j \in A} a_j\tag{9.7}$$

and $\mathcal{F} = \{A \in 2^E | w(A) \leqslant b\}$, then the pair $S = (E, \mathcal{F})$ is the independence system corresponding to K. Let

$$c(A) = \sum_{j \in A} c_j.$$

The knapsack problem may be written as

$$v(K) = \max_{A \in \mathcal{F}} c(A) = \max_{A \in \mathcal{B}} c(A).\tag{K}$$

A *matroid* on E is an independence system S on E with an additional condition

(M3) for any two sets $A, B \in \mathcal{F}$ such that $|B| = |A| + 1$, there exists an element $e \in B - A$ such that $A \cup \{e\} \in \mathcal{F}$.

There are many equivalent definitions of a matroid. For our purposes we need the following one.

THEOREM 9.5. *A pair* $(E, \mathcal{F})$ *is a matroid if and only if it satisfies* (M1), (M2) *and*

(M3') *for any* $C \subseteq E$, *every two maximal independent subsets of* C *have the same cardinality.*

Proof. ($\Rightarrow$) Let $A, B \subseteq C$ be maximal independent subsets such that $|B| > |A|$. If we choose $B' \subseteq B$ such that $|B'| = |A| + 1$, by (M1) and (M2), $B' \in \mathcal{F}$, and by (M3), there exists an element $e \in B' - A$ such that $A \cup \{e\} \in \mathcal{F}$, which contradicts the maximality of A.

($\Leftarrow$) Assume now that (M1), (M2) and (M3') hold. Let $A, B \in \mathcal{F}$ be sets such that $|B| = |A| + 1$ and let $C = A \cup B$. We assume now that there is no element $e \in B - A$ such that $A \cup \{e\} \in \mathcal{F}$. This means that A is the maximal independent subset of C. If we eventually extend B to the maximal independent subset $B' \subseteq C$, we get $|A| < |B'|$, which contradicts (M3'). $\qquad \square$

The *rank* of a set $C \subset E$, denoted as $r(C)$, is the cardinality of a maximal independent subset of C. In other words,

$$r(C) = \max\{|A| | A \in \mathcal{F}, A \subseteq C\}.$$

So $r(C) = C$ if and only if $C \in \mathcal{F}$. From Theorem 9.5 we have

COROLLARY 9.1. *Every two bases of a matroid have the same cardinality.*   □

For instance, all bases of a linear space have the same cardinality.

It is interesting to apply the greedy algorithm to a matroid. Let $E = \{e_1, ..., e_n\}$ and let $f: E \to R^+$ be a given weight (cost) function which associates with every $e \in E$ its weight (cost) $c(e)$. For any $A \subseteq E$, a weight of A is defined similarly as (9.7).

For a given matroid $M = (E, \mathscr{F})$, we will consider the system

$$v(P) = \max_{A \in \mathscr{F}} c(A). \qquad (P)$$

Without loss of generality, we assume that

$$c_1(e_1) \geqslant c_2(e_2) \geqslant ... \geqslant c(e_n). \qquad (9.8)$$

The Greedy Algorithm for the Matroid

Step 0 (Initialization): $A := \varnothing$;

Step 1 (Basic Iteration): For $i = 1, 2, ..., n$, if $A \cup \{e_i\} \in \mathscr{F}$, then $A := A \cup \{e_i\}$; Stop.   □

THEOREM 9.6. *If $M = (E, \mathscr{F})$ is a matroid, then a set A^* constructed by the greedy algorithm is an optimal solution to P ($A^* \in F^*(P)$), otherwise, if M is not a matroid, then there exists a function $f: E \to R^+$ such that $A^* \notin F^*(P)$.*

Proof. Let $M = (E, \mathscr{F})$ is a matroid and let $A^* = \{a_1, ..., a_k\}$. We have to prove that for any $T = \{t_1, ..., t_m\} \in \mathscr{F}$, we have $c(T) \leqslant c(A^*)$. Observe that $m \leqslant k$ as A^* is a basis of M. We show that $c(T) \leqslant c(A^*)$ by proving that $c(t_i) \leqslant c(a_i)$ for all $i \leqslant m$. Assume the opposite, $(c(t_i) > c(a_i))$, and consider two independent subsets $A' = \{a_1, ..., a_{i-1}\}$ and $T' = \{t_1, ..., t_{i-1}, t_i\}$. Then, by (M3), there exists an element t_j with $j \leqslant i$ such that $\{a_1, ..., a_{i-1}, t_j\} \in \mathscr{F}$. Since $c(t_j) \geqslant c(t_i) > c(a_i)$, there exists an index $p \leqslant i$ such that $c(a_1) \geqslant c(a_2) \geqslant ... \geqslant c(a_{p-1}) \geqslant c(t_j) > c(a_j)$, which contradicts the principle saying that the greedy algorithm adds to $a_1, ..., a_{p-1}$ an element a_p with the largest weight which preserves independence. So $c(T) \leqslant c(A)$.

Assume now that $M = (E, \mathscr{F})$ is not a matroid. If condition (M2) does not hold, there exists two sets $A, B \subseteq E$ such that $A \subseteq B$, $B \in \mathscr{F}$, but $A \notin \mathscr{F}$. If we define

$$c(e) = \begin{cases} 1 & \text{if} \quad e \in A, \\ 0 & \text{if} \quad e \in E - A, \end{cases}$$

then a set A^* constructed by the greedy algorithm does not contain A and thus $c(A^*) < c(B) = c(A)$.

If condition (M3) does not hold, but (M2) does, there exist two independent sets A and B such that $|A| = k$, $|B| = k+1$, and for every $e \in B - A$, the set $A \cup \{e\} \notin \mathscr{F}$. Let $p = |A \cap B|$ (obviously $p < k$) and let $0 < \varepsilon < 1/(k-p)$. If we define

$$c(e) = \begin{cases} 1+e & \text{if} \quad e \in A, \\ 1 & \text{if} \quad e \in B - A, \\ 0 & \text{otherwise,} \end{cases}$$

the greedy algorithm selects first all $e \in A$ and next deletes all $e \in B - A$. Thus

$$c(A^*) = c(A) = k(1+\varepsilon) = (k-p)(1+\varepsilon) + p(1+\varepsilon)$$

$$< (k-p)\frac{k+1-p}{k-p} + p(1+\varepsilon) = (k+1-p) + p(1+\varepsilon) = c(B). \qquad \square$$

So the greedy algorithm solves P in $O(n)$, providing that elements are ordered in such a way that (9.8) holds. It is interesting to note that in the construction of A^* the ordering (9.8) is important and the greedy algorithm does not take into account $c(e)$, $e \in E$. The consideration may be modified for the case of the problem

$$v(P) = \min_{A \in \mathscr{F}} c(A), \qquad (P)$$

where $M = (E, \mathscr{F})$ is a given matroid.

As an application of the above results, we consider a connected graph $G = (V, E)$ and define a matroid $M(G) = (E, \mathscr{F})$, where

$$\mathscr{F} = \{A \subseteq E | \text{graph}(V, A) \text{ does not contain cycles}\}.$$

THEOREM 9.7. $M(G)$ is a matroid.

Proof. The conditions (M1) and (M2) are obviously satisfied. The maximal independent sets of $M(G)$ are spanning trees of G and they all have $|V| - 1$ edges. So (M3') holds. $\qquad \square$

Consider the shortest spanning tree problem in a graph $G = (V, E)$. After ordering of edges, we have $c(e_1) \leqslant c(e_2) \leqslant \ldots \leqslant c(e_m)$, where $|E| = m$. It is easy to see that the greedy algorithm checks whether e_i forms a cycle with the so-far choosen edges $e_1, e_2, \ldots, e_{i-1}$.

COROLLARY 9.2. *The greedy algorithm finds a shortest tree in a given graph in* $O(|E|)$. $\qquad \square$

If a pair $(E, \mathscr{F})$ is only an independence system but not a matroid, the greedy algorithm gives a near-optimal solution to the problem

$$v(P) = \max_{A \in \mathscr{F}} c(A). \qquad (P)$$

Now we estimate the error of such a solution. The *lower rank* of a set C in a system $S = (E, \mathscr{F})$ is defined as

$$l(C) = \min \{|A| | A \in \mathscr{F}, A \subseteq C, \forall e \in E - A, A \cup \{e\} \notin \mathscr{F}\}.$$

In this context, $r(C)$ may be considered as the *upper rank* of C.

THEOREM 9.8. *Let* A' *be a near-optimal solution to* P *obtained by the greedy algorithm and let* $A^* \in F^*(P)$. *Then, for any function* $f \colon E \to R^+$,

$$\min_{X \subseteq E} \frac{l(X)}{r(X)} \leqslant \frac{c(A')}{c(A^*)} \leqslant 1. \qquad (9.9)$$

Proof. Let $E = \{e_1, \ldots, e_n\}$, $c(e_1) \geqslant c(e_2) \geqslant \ldots \geqslant c(e_n) \geqslant c(e_{n+1}) = 0$ and let $E_i = \{e_1, \ldots, e_i\}$. Then

$$c(A') = \sum_{i=1}^{n} |A' \cap E_i| \big(c(e_i) - c(e_{i+1}) \big), \tag{9.10}$$

$$c(A^*) = \sum_{i=1}^{n} |A^* \cap E_i| \big(c(e_i) - c(e_{i+1}) \big). \tag{9.11}$$

Since $A^* \cap E_i \in \mathscr{F}$, we have $|A^* \cap E_i| \leqslant r(A^* \cap E_i)$. The set $A' \cap E_i$ is the maximal independent subset of E_i. Therefore, $|A' \cap E_i| \geqslant l(E_i)$ and

$$|A' \cap E_i| \geqslant |A^* \cap E_i| \frac{l(E_i)}{r(E_i)} \geqslant |A^* \cap E_i| \min_{X \subseteq E} \frac{l(X)}{r(X)}.$$

Then, by (9.10) and (9.11), we get

$$\frac{c(A')}{c(A^*)} \geqslant \min_{X \subseteq E} \frac{l(X)}{r(X)} \leqslant 1. \qquad \square$$

As in any matroid we have $l(X) = r(X)$ for any $X \subseteq E$, we obtain a new, algorithmic definition of a matroid:

COROLLARY 9.3. *An independence system $(E, \mathscr{F})$ is a matroid if and only if, for any function $f: E \to R^+$, the greedy algorithm finds an optimal solution to the problem*

$$v(P) = \max_{A \in \mathscr{F}} c(A). \qquad \square$$

9.6. BIBLIOGRAPHIC NOTES

9.1. Fisher (1980) gives a review on near-optimal methods in integer programming. Karp in 1976 and in 1977 gave a description of the probabilistic analysis. Its application to the simple plant location problem may be found in Cornuejols et al. (1977) and in Fisher and Hochbaum (1980). An annotated bibliography on approximation algorithms is given in Garey and Johnson (1976).

9.2. Theorem 9.1 is proved in Kannan and Korte (1978). Walukiewicz (1975) proved Theorem 9.2 and presented results of computational experiments. For the so-called practical knapsack problem, the greedy algorithm gives solutions with a relative error not greater than 1%.

9.3. The basic references are Sahni (1977) and Sahni and Horowitz (1978). Theorem 9.3 is proved in Sahni (1975). Polynomial approximation schemes for the knapsack problem are described in Ibarra and Kim (1975) and in Lawler (1977). *FAS*'s for the simple plant location problem are considered in Fisher et al. (1978) and in Nemhauser et al. (1978).

9.4. The estimation (9.6) is given in Christofides (1976b) (see also Cornuejols and Nemhauser, 1978). An interesting modification of local methods, the so-called thermodynamical approach, is described in Kirkpatrick et al. (1982) and in Burkard and Rendl (1983).

9.5. Whitney defined a matroid in 1935 (see also Korte, 1981). Theorem 9.6 was proved independently by Rado in 1957 and by Edmonds in 1971. The greedy algorithm for the shortest spanning tree problem is described in Kruskal (1956). Theorem 9.8 is proved in Jenkyns (1976) and in Korte and Hausmann (1978).

9.7. EXERCISES

9.7.1. Prove Theorem 9.2.

9.7.2. Give a formal description of Algorithm A6.

9.7.3. Solve the problem from Exercise 5.8.2 by Algorithm A6.

9.7.4. Solve the above-mentioned problem by the rounding method.

9.7.5. Prove a theorem similar to Theorem 9.4 for the quadratic assignment problem (Sahni and Horowitz, 1978).

9.7.6. A vector $\hat{x}$ is called a Δ-solution to K if for given $\Delta > 0$, $v(t) - c\hat{x} \leqslant \Delta$ for any $t \in K$. Prove that finding a Δ-solution to K is $\mathcal{NP}$-hard (Sahni and Horowitz, 1978).

9.7.7. Demonstrate that in any tree the number of nodes with odd degrees is even.

9.7.8. Show that estimation (9.6) is tight.

9.7.9. Prove the following properties of the rank function for any sets $A, B \subseteq E$:
(a) $0 \leqslant r(A) \leqslant |A|$;
(b) if $A \subseteq B$, then $r(A) \leqslant r(B)$;
(c) $r(A \cup B) + r(A \cap B) \leqslant r(A) + r(B)$.

9.7.10. Prove that for any two circuits C, D of a matroid $M = (E, \mathcal{F})$ we have
(a) if $C \subseteq D$, then $C = D$;
(b) if $C \neq D$ and $e \in C \cap D$, then there exists a circuit $H \subset (C \cup D) - \{e\}$;
(c) if $A \in \mathcal{F}$, then, for any $e \in E$, the set $A \cup \{e\}$ contains at most one circuit.

9.7.11. Estimate ε in Exercise 9.7.3 using Theorem 9.8.

9.7.12. Prove that estimation (9.9) is tight.

9.7.13. Construct an independence system for the node packing problem and apply Theorem 9.8.

CHAPTER 10

Conclusions

The rapid development of integer programming in the last 10–15 years may be measured not only by the increasing number of publications but also by increasing abilities of computer codes. Inasmuch as at the end of the sixties, practical integer programming problems were commonly considered intractable, now, in the mid-eighties, it is difficult to find a practical problem, not an artificially constructed one, which cannot be solved at least near-optimally by existing codes.

The aim of this short chapter is to give a general description of computer codes for solving integer programming problems and to discuss some future trends in integer programming.

10.1. COMPUTER CODES

By a system (of computer codes) we mean a set of interrelated procedures which enable input of data, output of results and control of the solution process. Usually, it is not assumed that the user knows numerical methods used in the procedures although such knowledge is helpful.

Garfinkel (1979) gives a description of 11 firm systems (packages) for solving mixed integer programming problems. All systems are based on the branch-and-bound approach (see Chapter 4). Generally speaking, there are two ways of solving a given integer problem by a system. First, automatically, without any help from the user, and second with contribution by the user. In the first approach, a given problem is solved in a fixed way, e.g., after solving the linear programming relaxation, the first fractional variable is chosen as a branching variable, the first subproblem on the list of subproblems is chosen for analysis, and so on.

Some computer codes contain reasonable options for controlling the enumeration. One such option is a stopping rule for discounting the process prior to verification of optimality. Another option is to stop when a solution is found that is within a given percentage of $v(\bar{P})$. It may also be possible to stop after a prespecified number of integer solutions have been located on the tree. This option, although not as intuitively appealing as the others, is based on the supposition that the best solution tends to be found early and that the latter stages of enumeration are mainly spent in verifying optimality. This has often been found to be the case in practice. The user may also assign priorities to variables, change the rules of choosing a subproblem from the list or start enumeration from a given feasible solution. Usually, the systems allow some modifications of data, e.g., changing the right-hand-side vector or some coefficients in the objective function or in the constraint matrix.

The output of the results may be more or less detailed. For instance, the user may require the output of information about all incumbents obtained in the enumeration. Such information is useful in many practical situations when we are not only interested in the optimal solution but also in some feasible solutions close to the optimal one.

It should be pointed out that solving a given integer problem by a given system is, generally, not a single act. Usually, we have to solve some number of problems, for instance, for different cost coefficients, different right-hand-side vectors, etc. Then the first computations are done automatically, while in the next computations, using the information gained about the problem, we may specify the priorities of variables, start computations from known feasible solutions, change the output of results, and so on.

The system called PIPEX (Pure Integer Programming/Executor) opens a new generation of computer codes for solving integer programming problem. The system was developed at IBM and is described in detail in Crowder et al. (1983). It solves large (pure) integer programming problems and the computation process may be divided into three phases:

(1) preprocessing phase,
(2) constraint generation phase,
(3) branch-and-bound phase.

The automatic problem preprocessing essentially consists of "inspecting" the user-supplied formulation of a binary problem by tightening the formulation, by fixing some variables at their optimal values and/or by determining constraints which are inactive by the previous data manipulations (see Chapter 6 for details).

The constraint generation is done according to the principles described in Chapter 6. Such additional constraints are facets of corresponding knapsack polytopes and they preserve the sparsity of the constraint matrix.

After the end of the second phase, we have a tighter equivalent formulation of a given binary problem which is solved by branch-and-bound using the IBM package called the MPSX/370 (Mathematical Programming System Extended/370) and in particular its part called the MIP/370 (Mixed Integer Programming/370). So PIPEX may be considered an extension of MIP/370.

The paper by Crowder et al. (1983) contains results of computational experiments with PIPEX. Ten large problems taken from practice were solved up to optimality: the largest of them had 2756 variables and 756 constraints. This problem was solved (and the optimality was proved) for the first time by PIPEX within 54.42 CPU minutes of the IBM 370/168 computer. It is interesting to note that the first two phases took only 5.10 minutes and reduced the duality gap by 98%.

Recently, due to increasing popularity of microcomputers and personal computers in particular, significant progress has been made in optimization software. As a result, we have any optimization systems, many of them can be run on personal computers such as IBM PC/AT or their equivalents. Almost all scientific journals have a software reviews which provide the latest information about optimization software.

10.2. FUTURE TRENDS

Finally, we would like to point out some future trends in integer programming.

1. *From detail to the whole.* Study of particular integer programming problems gives results which may be used in the construction of methods for solving general problems. A good example is here constraint generation used in PIPEX. It is reasonable to assume that, for instance, facets of the multiple-choice knapsack polytope will be of great help in solving large integer programming problems with special ordered sets constraints.

2. *Hybrid algorithms.* In the case when we know many numerical methods for a given problem, it is reasonable to combine some of them into a hybrid algorithm, which in the course of computations, depending on the values of some parameters, will automatically choose one of them.

3. *Analysis of integer programming algorithms* will be intensively carried out in the coming years for at least two reasons. First, as a result of it, we have more efficient algorithms based on new results or new data structures. Second, as parallel computations are becoming more and more popular, we now have to analyse sequential algorithms in order to construct their parallel version or new parallel algorithms.

4. *Integration of different parts of mathematical programming.* Even at the beginning of the seventies, the relations between integer and nonlinear programming were quite weak. Held and Karp (1970), who presented a successful application of subgradient methods for solving the travelling salesman problem, initiated a heavy stream of papers devoted to application of nonlinear programming methods in integer programming. The ellipsoid method and Karmarkar's method are more recent examples of such integration. Obviously, increasing the efficiency of these nonlinear programming methods results in more efficient integer programming methods, e.g., branch-and-bound methods. On the other hand, efficient methods for solving linear multiple-choice knapsack problems are contributions to linear programming. One should expect further integration of these parts of mathematical programming in the coming future.

Bibliography

Aho, A. V., Hopcroft, J. E. and Ullman, J. D.: (1974), *The Design and Analysis of Computer Algorithms*, Addison–Wesley, Reading, London.

Armstrong, R. O., Sinha, P. and Zoltners, A. A.: (1982), 'The Multiple-Choice Nested Knapsack Model', *Management Sci.* **28**, 34–43.

Babat, L. G.: (1979), 'Growth of Coefficients in Integral Linear Programming', *Mathematical Methods of Solution of Economic Problems* **8**, 34–43 (in Russian).

Balas, E.: (1975), 'Facets of the Knapsack Polytope', *Math. Prog.* **8**, 146–164.

Balas, E. and Bergthaller, C.: (1983), 'Benders' Method Revisted', *J. Comp. and Appl. Math.* **9**, 3–12.

Balas, E. and Christofides, N.: (1981), 'A Restricted Lagrangean Approach to the Travelling Salesman Problem', *Math. Prog.* **21**, 19–46.

Balas, E. and Padberg, M.: (1976), 'Set Partitioning: A Survey', *SIAM Rev*, **18**, 710–760.

Balas, E. and Zemel, E.: (1978), 'Facets of the Knapsack Polytope from Minimal Covers', *SIAM J. Appl. Math.*, 119–148.

Balas, E. and Zemel, E.: (1980), 'An Algorithm for Large Zero–One Knapsack Problems', *Opns. Res.* **28**, 1130–1154.

Baliński, M. L.: (1965), 'Integer Programming: Methods, Uses, Computation', *Man. Sci.* **12**, 253–313.

Baliński, M. L. and Spielberg, K.: (1969), 'Methods for Integer Programming: Algebraic, Combinatorial and Enumerative', in: Aronofsky, J. (ed.): (1969), *Progress of Operations Research*, John Wiley and Sons, London, Vol. 3, 195–292.

Bazaraa, M. S. and Goode, J. J.: (1977), 'The Travelling Salesman Problem: A Duality Approach', *Math. Prog.* **13**, 221–237.

Bednarczuk, E.: (1977), 'On the Results of Solving some Linear Programming Problems Using Programs Packages of IBM and ROBOTRON Computers', *Prace IBS PAN*, No. 2 (in Polish).

Bell, D. E. and Shapiro, J. F.: (1977), 'A Convergent Duality Theory for Integer Programming', *Opns. Res.* **25**, 419–434.

Bellman, R.: (1957), *Dynamic Programming*, Princeton University Press.

Bellmore, M. and Malone, J. C.: (1971), 'Pathology of Travelling-Salesman Subtour Elimination Algorithms', *Opns. Res.* **19**, 278–307.

Bellmore, M. and Nemhauser, G. L.: (1968), 'The Travelling Salesman Problem: A Survey', *Opns. Res.* **16**, 538–558.

Benders, J. F.: (1962), 'Partitioning Procedures for Solving Mixed-Variables Programming Problems', *Numer. Math.* **4**, 238–252.

Berge, C.: (1962), *The Theory of Graphs and its Application*, Methuen, translated by A. Doig from the original French edition, Dunod, 1958.

Blair, C. E. and Jeroslaw, R. G.: (1977), 'The Value Function of a Mixed Integer Program, I', *Discrete Math.* **19**, 121–138.

Blair, C. E. and Jeroslaw, R. G.: (1979), 'The Value Function of a Mixed Integer Program, II', *Discrete Math.* **25**, 7–19.

Bland, R. G.: (1977), 'New Finite Pivoting Rules for the Simplex Method', *Math. Opns. Res.* **2**, 103–107.

Bland, R. G., Goldfarb, D. and Todd, M. J.: (1981), 'The Ellipsoid Method: A Survey', *Opns. Res.* **29**, 1039–1091.

Bowman, V. J. and Nemhauser, G. L.: (1971), 'Deep Cuts in Integer Programming', *Opns. Res.* **8**, 89–111.

Bradley, G. H.: (1971), 'Equivalent Integer Programs and Canonical Problems', *Management Sci.* **17**, 354–366.

Bradley, G. H., Hammer, P. L. and Wolsey, L. A.: (1974), 'Coefficient Reduction for Inequalities in 0–1 Variables', *Math. Prog.* **7**, 263–282.

Breu, R. and Burdet, C. A.: (1974), 'Branch and Bound Experiments in 0–1 Programming', *Math. Prog. Study* **2**, 1–50.

Burkard, R. E.: (1979), 'Travelling Salesman and Assignment Problems: A Survey', *Ann. Discrete Math.* **4**, 193–215.

Burkard, R. E., Hahn, W. and Zimmermann, U.: (1977), 'An Algebraic Approach to Assignment Problems', *Math. Prog.* **12**, 318–327.

Burkard, R. E. and Rendl, F.: (1983), 'A Thermodynamically Motivated Simulation Procedure for Combinatorial Optimization Problems', *European J. Opns. Res.* **17**, 169–174.

Chalmet, L. G. and Gelders, L. F.: (1976), 'Lagrangean Relaxations for Generalized Assignment-Type Problem', *Proceedings of the Second European Congress on Operations Research*, North-Holland, Amsterdam, New York, 103–110.

Chandrasekaran R.: (1969), 'Total Unimodularity of Matrices', *SIAM J.* **11**, 1032–1034.

Charnes, A., Cooper, W. W., Duffuaa, S. and Kress, M.: (1980), 'Complexity and Computability of Solutions to Linear Programming Systems', *Internat. J. Computer Infor. Sci.* **9**, 483–506.

Christofides, N.: (1972), 'Bounds for the Travelling Salesman Problem', *Opns. Res.* **20**, 1044–1056.

Christofides, N.: (1976a), 'The Vehicle Routing Problem', *Cahiers Centre Études Recherche Oper.* **10**, 55–70.

Christofides, N.: (1976b), 'Worst-Case Analysis of a New Heuristic for the Travelling Salesman Problem', Carnegie Mellon U. Report WP 62-75-76, Pittsburgh.

Chvatal, V.: (1983), *Linear Programming*, Freeman, New York.

Chvatal, V. and Hammer, P. L.: (1977), 'Aggregation of Inequalities in Integer Programming', *Ann. Discrete Math.* **1**, 145–162.

Clarke, G. and Wright, J.: (1964), 'Scheduling of Vehicles from a Central Depot to a Number of Delivery Points" *Opns. Res.* **12**, 568–581.

Clausen, J.: (1980), 'A Tutorial Note on the Complexity of the Simplex-Algorithm', in: Krarup, J. and Walukiewicz, S. (eds.): (1980), *Proceedings of DAPS-79*, Institute of Datalogy, University of Copenhagen, 51–65.

Cook, S. A.: (1971), 'The Complexity of the Theorem-Proving Procedures', *Proceedings 3rd Annual ACM Symposium on Theory of Computing*, Cook, New York, 151–158.

Cornuejols, G., Fisher, M. L. and Nemhauser, G. L.: (1977), 'Location of Bank Accounts to Optimize Float: An Analytic Study of Exact and Approximate Algorithms', *Management Sci.* **23**, 789–810.

Cornuejols, G. and Nemhauser, G. L.: (1978), 'Thight Bound for Christofides' Travelling Salesman Heuristic' *Math. Prog.* **14**, 116–121.

Crowder, H., Johnson, E. L. and Padberg, M.: (1983), 'Solving Large-Scale Zero–One Linear Programming Problems', *Opns. Res.* **31**, 803–834.

Dantzig, G. B.: (1957), 'Discrete Variable Extremum Problems', *Opns. Res.* **5**, 266–277.

Dantzig, G. B.: (1963), *Linear Programming and Extensions*, Princeton University Press.

Dantzig, G. B., Fulkerson, D. R. and Johnson, S. M.: (1954), 'Solution of a Large-Scale Travelling-Salesman Problem', *Opns. Res.* **2**, 393–410.

Dembo, R. S. and Hammer, P. L.: (1980), 'A Reduction Algorithm for Knapsack Problems', *Methods Opns, Res.* **36**, 49–60.

Deo, N.: (1974), *Graph Theory with Applications to Engineering and Computer Science*, Prentice Hall, Englewood Cliffs.

Derigs, U. and Zimmermann, U.: (1978), 'An Augmenting Path Method for Solving Linear Bottleneck Assignment Problems', *Computing* **19**, 285–295.

Dijkstra, E.: (1959), 'A Note on Two Problems in Connection with Graphs', *Numer. Math.* **1**, 269–271.

Driebeek, N. J.: (1966), 'An Algorithm for the Solution of Mixed Integer Programming Problems', *Management Sci.* **12**, 576–587.

Dudziński, K. and Walukiewicz, S.: (1984), 'A Fast Algorithm for Solving Linear Multiple-Choice Knapsack Problem', *Opns. Res. Lett.* **3**, 205–209.

Dudziński, K. and Walukiewicz, S.: (1987), 'Exact Methods for the Knapsack Problem and its Generalizations', *European J. Opns. Res.* **28**, 3–21.

Dyer, M. E.: (1980), 'Calculating Surrogate Contraints', *Math. Prog.* **19**, 255–278.

Dyer, M. E.: (1984), 'An $O(n)$ Algorithm for the Multiple-Choice Knapsack Linear Program', *Math. Prog.* **29**, 57–63.

Edmonds, J.: (1965a), 'Paths, Trees and Flowers', *Can. J. Math.* **17**, 449–467.

Edmonds, J.: (1965b), 'Maximum Matching and a Polyhedron with 0, 1 Verices', *J. Res. Nat. Bur. Stds.* **69B**, 125–130.

Edmonds, J.: (1971), 'Matroids and the Greedy Algorithm', *Math. Prog.* **1**, 127–136.

Etcheberry, J.: (1977), 'The Set-Covering Problem: A New Implicit Enumeration Algorithm', *Opns. Res.* **25**, 760–772.

Faner, M., Sosiński, R. and Walukiewicz, S.: (1981), 'Decomposition of Mixed 0–1 Programs with Nonlinear Discrete Part Using Pseudo-Boolean Programming', in: Walukiewicz, S. and Wierzbicki, A. (eds.): (1981), *Methods of Mathematical Programming*, PWN, Warsaw, 69–74.

Fayard, D. and Plateau, G.: (1982), 'An Algorithm for the Solution of the 0–1 Knapsack Problem', *Computing* **28**, 269–287.

Fisher, M. L.: (1980), 'Worst-Case Analysis of Heuristic Algorithms', *Management Sci.* **26**, 1–18.

Fisher, M. L.: (1981), 'Lagrangian Relaxation Method for Solving Integer Programming Problems', *Management Sci.* **27**, 1–18.

Fisher, M. L. and Hochbaum, D. S.: (1980), 'Probabilistic Analysis of the Planar k-Median Problem', *Math. Opns. Res.* **5**, 27–34.

Fisher, M. L., Nemhauser, G. L. and Wolsey, L. A.: (1978), 'An Analysis of Approximations for Maximizing Submodular Set Functions—II', *Math. Prog. Study* **8**, 73–87.

Ford, L. R., and Fulkerson, D. R.: (1962), *Flows in Networks*, Princeton University Press.

Forrest, J. J. H., Hirst, J. P. H. and Tomlin, J. A.: (1974), 'Practical Solution of Large Mixed Integer Programming Problems with EMPIRE', *Management Sci.* **20**, 736–773.

Frederickson, G. N. and Johnson, D. B.: (1982), 'The Complexity of Selection and Ranking in $X+Y$ and Matrices with Sorted Columns', *J. Computer and System Sci.* **24**, 197–208.

Garey, M. R. and Johnson, D. S.: (1976), 'Approximation Algorithms for Combinatorial Problems: An Annotated Bibliography', in: Traub, J. F. (ed.): (1976), *Algorithms and Complexity. New Directions and Recent Results*, Academic Press, New York, 41–52.

Garey, M. R. and Johnson, D. S.: (1979), *Computers and Intractability: A Guide to the Theory of NP-Completeness*, Freeman, San Francisco.

Garfinkel, R. S.: (1979), 'Branch-and-Bound Method for Integer Programming', in: Christofides, N., Mingozzi, A., Toth, P. and Sandi, C. (eds.): (1979), *Combinatorial Optimization*, John Wiley and Sons, New York, 1–20.

Garfinkel, R. S. and Nemhauser, G. L.: (1972), *Integer Programming*, John Wiley and Sons, London.

Garfinkel, R. S. and Nemhauser, G. L.: (1973), 'A Survey of Integer Programming Emphasizing Computation and Relations among Models', in: Hu, T. C. and Robinson, S. M. (eds.): (1973), *Mathematical Programming*, Academic Press, New York, 77–155.

Gass, S. I.: (1964), *Linear Programming Methods and Applications*, 2nd ed., McGraw-Hill, New York.

Geoffrion, A. M.: (1969), 'An Improved Implicit Enumeration Approach for Integer Programming, *Opns. Res.* **17**, 437–454.

Geoffrion, A. M.: (1972), 'Generalized Benders' Decomposition', *J. Optimization Theory and Appl.* **10**, 237–260.

Geoffrion, A. M.: (1974), 'Lagrangean Relaxation for Integer Programming', *Math. Prog. Study* **2**, 82–114.

Geoffrion, A. M. and Graves, G. W.: (1974), 'Multicommodity Distribution System Design by Benders' Decomposition', *Management Sci.* **20**, 822–844.

Geoffrion, A. M. and Marsten, R. E.: (1972), 'Integer Programming: A Framework an State-of-the-Art Survey', *Management Sci.* **18**, 465–491.

Geoffrion, A. M. and Nauss, R.: (1977), 'Parametric and Post-optimality Analysis in Integer Programming', *Management Sci.* **23**, 453–466.

Gill, P. E., Murray, W., Saunders, M. A. and Wright, M. H.: (1982), 'A Numerical Investigation of Ellipsoid Algorithm for Large-Scale Linear Programming', in: Dantzig, G. B., Dempster, M. A. H. and Kalio, M. J. (eds.): (1982), *Large-Scale Linear Programming*, IIASA, Laxenburg (Austria), 487–509.

Gilmore, P. C. and Gomory, R. E.: (1961), 'A Linear Programming Approach to the Cutting Stock Problem', *Opns. Res.* **9**, 849–859.

Gilmore, P. C. and Gomory, R. E.: (1963), 'A Linear Programming Approach to the Cutting Stock Problem', Pt. II, *Opns. Res.* **11**, 863–888.

Gilmore, P. C. and Gomory, R. E.: (1965), 'Multistage Cutting Stock Problems of Two and More Dimensions', *Opns. Res.* **13**, 94–120.

Gilmore, P. C. and Gomory, R. E.: (1966), 'The Theory of Computation of Knapsack Function', *Opns. Res.* **14**, 1045–1074.

Glover, F.: (1965a), 'A Bound Escalation Method for the Solution of Integer Linear Problems', *Cahiers Centre d'Études Recherche Oper.* **6**, 131–168.

Glover, F.: (1965b), 'A Hybrid-Dual Integer Programming Algorithm', *Cahiers Centre d'Études Recherche Oper.* **7**, 5–23.

Glover, F.: (1965c), 'A Multiphase-Dual Algorithm for the Zero–One Integer Programming Problem', *Opns. Res.* **13**, 879–919.

Glover, F.: (1967), 'A Pseudo Primal-Dual Integer Programming Algorithm', *J. Res. Nat. Bur. Stds.* **71B**, 187–195.

Glover, F.: (1968a), 'A New Foundation for a Simplified Primal Integer Programming Algorithm', *Opns. Res.* **16**, 727–740.

Glover, F.: (1968b), 'Surrogate Constraints', *Opns. Res.* **16**, 741–749.

Glover, F.: (1975), 'Surrogate Constraint Duality in Mathematical Programming', *Opns. Res.* **23**, 434–451.

Glover, F. and Klingman, D.: (1979), 'An $O(n\log n)$ Algorithm for LP Knapsack with GUB Constraints', *Math. Prog.* **17**, 345–361.

Goffin, J.-L.: (1979), 'Acceleration in the Relaxation Method for Linear Inequalities and Subgradient Optimization', Working Paper 79-10, Faculty of Management, McGill University, Montreal.

Goldfarb, D. and Reid, J. K.: (1977), 'A Practicable Steepest-Edge Simplex Algorithm', *Math. Prog.* **12**, 361–371.

Goldfarb, D. and Todd, M. J.: (1982), 'Modifications and Implementations of the Ellipsoid Algorithm for Linear Programming', *Math. Prog.* **23**, 1–19.

Gomory, R. E.: (1958), 'Outline of an Algorithm for Integer Solutions to Linear Programs', *Bull. Amer. Math. Soc.* **64**, 275–278.

Gomory, R. E.: (1963a), 'An Algorithm for Integer Solutions to Linear Programs', in: Graves, R. L. and Wolfe, P. (eds.): (1963), *Recent Advances in Mathematical Programming*, McGraw-Hill, New York, 269–302.

Gomory, R. E.: (1963b), 'All-Integer Integer Programming Algorithm', in: Muth, J. F. and Thompson, G. L. (eds.): (1963), *Industrial Scheduling*, Prentice-Hall, Englewoods Cliffs, 193–206.

Gomory, R. E.: (1965), 'On the Relation Between Integer and Non-Integer Solutions to Linear Programs', *Proc. Nat. Acad. Sci.* **53**, 260–265.

Gondran, M. and Minoux, M.: (1979), *Graphes et Algorithmes*, Editions Eyrolles, Paris.

Gould, F. J. and Rubin, D. S.: (1973), 'Rationalizing Discrete Programs', *Opns. Res.* **21**, 343–345.

Greenberg, H. J. and Pierskalla, W. P.: (1970), 'Surrogate Mathematical Programming', *Opns. Res.* **18**, 924–939.

Greenberg, H. J. and Pierskalla, W. P.: (1971), 'A Review of Quasi-Convex Functions', *Opns. Res.* **19**, 1553–1570.

Grötschel, M., Lovasz, L. and Schrijver, A.: (1982), 'The Ellipsoid Method and its Consequences in Combinatorial Optimization', *Combinatorica* **1**, 169–197.

Hammer, P. L. and Rudeanu, S.: (1968), *Boolean Methods in Operations Research and Related Areas*, Springer-Verlag, Berlin.

Hammer, P. L., Johnson, E. L. and Peld, U. N.: (1975), 'Facets of Regular 0–1 Polytopes', *Math. Prog.* **8**, 179–206.

Hansen, K. H. and Krarup, J.: (1974), 'Improvements for the Held–Karp Algorithm for the Symmetric Travelling-Salesman Problem', *Math. Prog.* **4**, 87–96.

Harary, F.: (1969), *Graph Theory*, Addison–Wesley, New York.

Hausmann, D. (ed.): (1978), 'Integer Programming and Related Areas, A Classified Bibliography', *Lecture Notes in Economics and Mathematical Systems*, Vol. 160, Springer-Verlag, Berlin.

Held, M. and Karp, R. M.: (1970), 'The Travelling-Salesman Problem and Minimum Spanning Trees', *Opns. Res.* **18**, 1138–1162.

Held, M. and Karp, R. M.: (1971), 'The Travelling-Salesman Problem and Minimum Spanning Trees', Pt. II, *Math. Prog.* **1**, 6–25.

Held, M., Wolfe, P. and Crowder, H. D.: (1974), 'Validation of Subgradient Optimization', *Math. Prog.* **6**, 62–88.

Hoffman, A. J. and Kruskal, J. B.: (1958), 'Integral Boundary Points of Convex Polyhedra', in: Kuhn, H. W. and Tucker, A. W. (eds.): (1958), *Linear Inequalities and Related Systems*, Princeton University Press, Princeton, 223–246.

Hu, T. C.: (1970), 'On the Asymptotic Integer Algorithm', *Linear Algebra and its Appl.* **3**, 279–294.

Ibaraki, T., Hasegawa, T., Teranaka, K. and Iwase, J.: (1978), 'The Multiple-Choice Knapsack Problem', *J. Opns. Res. Society of Japan* **21**, 59–94.

Ibarra, O. H. and Kim, C. E.: (1975), 'Fast Approximation Algorithm for the Knapsack and Sum of Subset Problem', *Journal of ACM* **22**, 463–468.

Ingargiola, G. P. and Korsh, J. F.: (1973), 'Reduction Algorithm for Zero–One Single Knapsack Problems', *Management Sci.* **20**, 460–463.

Ingargiola, G. P. and Korsh, J. F.: (1977), 'General Algorithm for One-Dimensional Knapsack Problems', *Opns. Res.* **25**, 752–759.

Jenkyns, T. A.: (1976), 'The Efficiency of the "Greedy" Algorithm', *Proc. 7th S.-E. Conf. Combinatorics, Graph Theory and Computing*, Winnipeg, 341–350.

Jeroslow, R. G.: (1977), 'Cutting-Plane Theory: Disjunctive Methods', *Ann. Discrete Math.* **1**, 293–330.

Jeroslow, R. G. (1978), 'Cutting-Plane Theory: Algebraic Methods', *Discrete Math.*, **23**, 121–151.

Johnson, D. B. and Mizogushi, T.: (1978), 'Selecting the k-th Element in $X+Y$ and $X_1+X_2+\dots \dots +X_n$', *SIAM Journal on Computing* **7**, 147–153.

Johnson, E. L. and Padberg, M.: (1981), 'A Note on the Knapsack Problem with Special Ordered Sets', *Opns. Res. Lett.* **1**, 18–22.

Kaliszewski, I. and Libura, M.: (1981), 'Constraints Aggregation in Integer Programming', in: Walukiewicz, S. and Wierzbicki, A. (eds.): (1981) *Methods of Mathematical Programming*, PWN, Warsaw, 161–167.

Kaliszewski, I. and Walukiewicz, S.: (1979), 'Tighter Equivalent Formulations of Integer Programming Problems', in: Prekopa, A. (ed.): (1979), *Survey of Mathematical Programming*, Akademiai Kiado, Budapest, 525–536.

Kaliszewski, I. and Walukiewicz, S.: (1981), 'A Transformation of Integer Programs', Report ZPM-2/81, Systems Research Institute, Warsaw.

Kannan, R. and Korte, B.: (1978), 'Approximative Combinatorial Algorithms', Report No. 78107--OR, University of Bonn, Bonn.

Karmarkar, N.: (1984), 'A New Polynomial-Time Algorithm for Linear Programming', Report AT and T Bell Lab., Murray Hill, N. J.

Karp, R. M.: (1972), 'Reducibility among Combinatorial Problems', in: Miller, R. E. and Thatcher, J. W. (eds.): (1972), *Complexity of Computer Computations*, Plenum Press, New York, 85–103,

Karp, R. M.: (1975), 'On Computational Complexity of Combinatorial Problems', *Networks* **5**, 45–68.

Karp, R. M.: (1976), 'The Probabilistic Analysis of Some Combinatorial Search Algorithms' in: Traub, J. F. (ed.): (1976), *Proc. Symp. on New Directions and Recent Results in Algorithms and Complexity*, Academic Press, N.Y., 1–19.

Karp, R. M.: (1977), 'Probabilistic Analysis of Partitioning Algorithms for the Travelling Salesman Problem in the Plane', *Math. Opns. Res.* **2**, 209–224.

Karwan, M. H. and Rardin, R. L.: (1979), 'Some Relationships between Lagrangian and Surrogate Duality in Integer Programming', *Math. Prog.* **17**, 320–334.

Kasting, C. (ed.): (1976), 'Integer Programming and Related Areas. A Classified Bibliography', *Lecture Notes in Economics and Mathematical Systems*, Vol. 128, Springer-Verlag, Berlin.

Khachian, L. G.: (1979), 'A Polynomial Algorithm for Linear Programming', *Doklady Akademii Nauk USSR* **244**, 1093–1096 (in Russian).

Kianfar, F.: (1971), 'Stronger Inequalities for 0–1 Integer Programming Using Knapsack Functions', *Opns. Res.* **19**, 1374–1392.

Kianfar, F.: (1976), 'Stronger Inequalities for 0–1 Integer Programming: Computational Refinements', *Opns. Res.* **24**, 581–585.

Kirpatrick, S., Gelatt, C. D. Jr. and Vecchi, M. P.: (1982), 'Optimization by Simulated Annealing', Research Report RC 9335, IBM Thomas J. Watson Research Center, Yorktown Heights, New York.

Klee, V. and Minty, G. J.: (1972), 'How Good is the Simplex Algorithm?', in: Shisha, O. (ed.): (1972), *Inequalities* III, Academic Press, New York, 159–175.

Kolokolov, A. A.: (1976), 'Upper Bounds for the Number of Cuts in Gomory Algorithm', in: *Metody Modelirovanya i Obrabotki Informacyi*, Nauka, Novosibirsk, 106–116 (in Russian).

Korte, B.: (1981), 'Matroids and Independence Systems', Report No. 79148-OR, University of Bonn.

Korte, B. and Hausman, D.: (1978), 'An Analysis of the Greedy Heuristic of Independence Systems', *Ann. Discrete Math.* **2**, 65–74.

Kotiah, T. C. T. and Steinberg, D. I.: (1978), 'On the Possibility of Cycling with the Simplex Method', *Opns. Res.* **26**, 374–376.

Krarup, J. and Pruzan, P. M.: (1983), 'The Simple Plant Location Problem: Survey and Synthesis', *European J. Opns. Res.* **12**, 36–81.

Kruskal, J. B.: (1956), 'On the Shortest Spanning Subtree of a Graph and the Travelling Salesman Problem', *Proc. Amer. Math. Soc.* **7**, 48–50.

Kuhn, H. W.: (1955), 'The Hungarian Method for Assignment Problems', *Nav. Res. Logistic Quart.* **2**, 83–97.

Land, A. H. and Doig, A. G.: (1960), 'An Automatic Method for Solving Discrete Programming Problems', *Econometrica* **28**, 497–520.

Land, A. H. and Powell, S.: (1979), 'Computer Codes for Problems of Integer Programming', *Ann. Discrete Math.* **5**, 221–269.

Lawler, E. L.: (1976), *Combinatorial Optimization: Networks and Matroids*, Holt, Rinehart and Winston, New York.

Lawler, E. L.: (1977), 'Fast Approximation Algorithms for Knapsack Problems', *Proc. 18th Annual Symposium of Foundations of Computer Science*, New York, 206–213.

Lawler, E. L. and Bell, M. D.: (1966), 'A Method for Solving Discrete Optimization Problems', *Opns. Res.* **14**, 1098–1112.

Lawler, E. L. and Wood, D. E.: (1966), 'Branch-and-Bound Methods, A Survey', *Opns. Res.* **14**, 699–719.

Lenstra, J. K., Rinnooy Kan, A. H. G. and Emde Boas van, P.: (1982), 'An Appraisal of Computational Complexity for Operations Researchers', *European J. Opns. Res.* **11**, 201–210.

Levin, A. J.: (1965), 'An Algorithm for Minimization of a Convex Function', *Doklady AN USSR* **160**, 1244–1247 (in Russian).

Libura, M.: (1977), 'Sensitivity Analysis of Solutions of the Integer Knapsack Problem', *Archiwum Automatyki i Telemechaniki* **22**, 313–322 (in Polish).

Little, J. D. C., Murty, K. G., Sweeney, D. W. and Karel, C.: (1963), 'An Algorithm for the Travelling Salesman Problem', *Opns. Res.* **11**, 972–989.

Marsten, R. E.: (1974), 'An Algorithm for Large Set Partitioning Problems', *Management Sci.* **20**, 774–787.

Marsten, R. E.: (1975), 'The Use of the Boxstep Method in Discrete Optimization', *Math. Prog. Study* **3**, 127–144.

Martello, S. and Toth, P.: (1978), 'Algorithm for the Solution of the 0–1 Single Knapsack Problem', *Computing* **21**, 81–86.

McDonald, D. and Devine, M.: (1977), 'A Modified Benders' Partitioning Algorithm for Mixed Integer Programming', *Management Sci.* **24**, 312–319.

Meyer, R. R.: (1974), 'On the Existence of Optimal Solutions to Integer and Mixed-Integer Programming Problems', *Math. Prog.* **7**, 223–235.

Miliotis, P.: (1978), 'Using Cutting Planes to Solve the Symmetric Traveling Salesman Problem', *Math. Prog.* **15**, 177–188.

Mitra, G.: (1973), 'Investigation of Some Branch and Bound Strategies for the Solution of Mixed Integer Linear Programs', *Math. Prog.* **4**, 155–170.

Mitten, L. G.: (1970), 'Branch-and-Bound Methods: General Formulation and Properties', *Opns. Res,* **18**, 24–34.

Motzkin, T. S. and Schonberg, I. J.: (1954), 'The Relaxation Method for Linear Inequalities', *Can. J. Math.* **6**, 393–404.

Nauss, R. M.: (1976), 'An Efficient Algorithm for the 0-1 Knapsack Problem', *Management Sci.* **23**, 27–31.

Nemhauser, G. L. and Trotter, L. E. Jr: (1974), 'Properties of Vertex Packing and Independence System Polyhedra', *Math. Prog.* **6**, 48–61.

Nemhauser, G. L., Wolsey, L. A. and Fisher, M. L.: (1978), 'An Analysis of Approximations for Maximizing Submodular Set Functions—I', *Math. Prog.* **14**, 265–294.

Padberg, M.: (1973), 'On Facial Structure of Set Packing Polyhedra', *Math. Prog.* **5**, 199–215.

Padberg, M.: (1979), 'Covering, Packing and Knapsack Problems', *Ann. Discrete Math.* **4**, 265–287.

Papadimitriou, C. H. and Steiglitz, K.: (1982), *Combinatorial Optimization: Algorithms and Complexity*, Prentice-Hall, Englewood Cliffs.

Pulleyblank, W. R.: (1983), 'Polyhedral Combinatorics' in: Bachem, A., Grotschel, M. and Korte, B. (eds.): (1983), *Mathematical Programming. The State of the Art*, Springer-Verlag, Berlin, 312–345.

Rado, R.: (1957), 'Note on Independence Function', *Proc. London Math. Soc.* **7**, 337–343.

Randow von, R. (ed.): (1982), 'Integer Programming and Related Areas. A Classified Bibliography 1978–1981', *Lecture Notes in Economics and Mathematical Systems*, Vol. 197, Springer-Verlag. Berlin.

Rosenberg, I. G.: (1974), 'Aggregation of Equations in Integer Programming', *Discrete Math.* **10**, 325–341.

Rubin, D. S. and Graves, R. L.: (1972), 'Strengthened Dantzig Cuts for Integer Programming', *Opns. Res.* **20**, 173–177.

Sahni, S.: (1975), 'Approximate Algorithms for 0/1 Knapsack Problems', *Journal of ACM* **22**, 115–124.

Sahni, S.: (1977), 'General Techniques for Combinatorial Approximation', *Opns. Res.* **25**, 920–936.

Sahni, S. and Horowitz, E.: (1978), 'Combinatorial Problems Reducibility and Approximation', *Opns. Res.* **26**, 718–759.

Salkin, H. M. and Breining, P.: (1971), 'Integer Points on Gomory Practitional Cut (Hyperplane)', *Nav. Res. Logistic Quart.* **18**, 491–496.

Schrijver, A.: (1986), *Theory of Linear and Integer Programming*, John Wiley and Sons, Chichester.

Shapiro, J. F.: (1968), 'Dynamic Programming Algorithms for the Integer Programming Problem—I: The Integer Programming Problem Viewed as a Knapsack Type Problem', *Opns. Res.* **16**, 103–121.

Shapiro, J. F.: (1977), 'Sensitivity Analysis in Integer Programming', *Ann. Discrete Math.* **1**, 467–477.

Shapiro, J. F.: (1979), *Mathematical Programming: Structures and Algorithms*, J. Wiley and Sons, New York.

Shor, N. Z.: (1985), *Minimization Methods for Non-Differemtiable Functions*, Springer-Verlag, Berlin, transl. from Russian.

Schönhage, A., Paterson, M. and Pippenger, N.: (1976), 'Finding the Median', *J. Computer Systems Sci.* **13**, 184–199.

Sinha, P. and Zoltners, A. A.: (1979), 'The Multiple-Choice Knapsack Problem', *Opns. Res.* **27**, 503–515.

Slisenko, A. O.: (1981), 'Complexity Problems in Theory of Computations', *Uspekhi Mat. Nauk* **36**, 21–103 (in Russian).

Słomiński, L.: (1976), 'An Efficient Approach to the Bottleneck Assignment Problem', *Bulletin of Polish Academy of Sciences, Ser. Sci. Tech.* **25**, 369–375.

Smale, S.: (1983), 'The Problem of the Average Speed of the Simplex Method', in: Bachem, A., Grotschel, M. and Korte, B. (eds.): (1983), *Mathematical Programming. The State of Art*, Springer-Verlag, Berlin, 530–539.

Sweeney, D. J. and Murphy, R. A.: (1979), 'A Method of Decomposition for Integer Programs', *Opns. Res.* **27**, 1128–1141.

Tind, J. and Wolsey, L. A.: (1981), 'An Elementary Survey of General Duality Theory in Mathematical Programming', *Math. Prog.* **21**, 241–261.

Truemper, K.: (1978), 'Algebraic Characterization of Unimodular Matrices', *SIAM J. App. Math.* **35**, 328–332.

Van Roy, T. J.: (1983), 'Cross Decomposition for Mixed Integer Programming', *Math. Prog.* **25**, 46–63.

Van Roy, T. J.: (1986), 'Cross Decomposition Algorithm for Capacited Facility Location', *Opns. Res.* **34**, 143–163.

Veinott, A. F. Jr. and Dantzig, G. B.: (1968), 'Integral Extreme Points', *SIAM Rev.* **10**, 371–372.

Waluk, B. and Walukiewicz, S.: (1981), 'Computational Experiment with the Ellipsoid Algorithm' *Prace IBS PAN* **76** (in Polish).

Walukiewicz, S.: (1975), 'Almost Linear Integer Programming Problems', *Prace IOK* **23** (in Polish).

Walukiewicz, S.: (1981), 'Some Aspects of Integer Programming Duality', *European J. Opns. Res.* **7**, 196–202.

Walukiewicz, S., Słonimski, L. and Faner, M.: (1973), 'An Improved Algorithm for Pseudo-Boolean Programming', in: Conti, R. and Ruberti, A. (eds.): (1973), *5th IFIP Conference on Optimization Techniques*, Pt. 1, Springer-Verlag, Berlin, 493–504.

Watters, L. J.: (1967), 'Reduction of Integer Polynomial Programming Problems to Zero–One Linear Programming Problems', *Opns. Res.* **15**, 1171–1174.

Whitney, H.: (1935), 'On the Abstract Properties of Linear Independence', *Amer. J. Math.* **57**, 509–533.

Williams, H. P.: (1974), 'Experiments in the Reformulation of Integer Programming Problems', *Math. Prog. Study* **2**, 180–197.

Williams, H. P.: (1978a), *Model Building in Mathematical Programming*, John Wiley and Sons, Chichester.

Williams, H. P.: (1978b), 'The Formulation of Two Mixed Integer Programming Problems', *Math. Prog.* **14**, 325–331.

Witzgal, C.: (1977), 'On One-Row Linear Programs', Applied Mathematics Division Report, National Bureau of Standards.

Wolfe, P.: (1980), 'A Bibliography for the Ellipsoid Algorithm', IBM Research Centre, Report No. 8237.

Wolsey, L. A.: (1976), 'Further Facets Generating Procedures for Vertex Packing Polytopes', *Math. Prog.* **11**, 158–163.

Wolsey, L. A.: (1981), 'Integer Programming Duality: Price Functions and Sensivity Analysis', *Math. Prog.* **20**, 173–195.

Young, R. D.: (1968), 'A Simplified Primal (All-Integer) Integer Programming Algorithm', *Opns. Res.* **16**, 750–782.

Yudin, D. B. and Nemirowskii, A. S.: (1976), 'Informational Complexity and Effective Methods of Solution for Convex Extremal Problems', *Ekonomika i Matematicheskie Metody* **12**, 337–369 (in Russian).

Zemel, E.: (1980), 'The Linear Multiple-Choice Knapsack Problem', *Opns. Res.* **28**, 1412–1423.

Zoltners, A. A.: (1978), 'A Direct Descent Binary Knapsack Algorithm', *Journal of ACM* **25**, 304–311.

Index